T0295465

SMART TECHNOLOGIES FOR SUSTAINABLE SMALLHOLDER AGRICULTURE

SMART TECHNOLOGIES FOR SUSTAINABLE SMALLHOLDER AGRICULTURE

Upscaling in Developing Countries

Edited by

NHAMO NHAMO

DAVID CHIKOYE

THERESE GONDWE
International Institute of Tropical Agriculture (IITA)
Southern Africa Research and Administration Hub
(SARAH) Campus, Lusaka, Zambia

Academic Press is an imprint of Elsevier
125 London Wall, London EC2Y 5AS, United Kingdom
525 B Street, Suite 1800, San Diego, CA 92101-4495, United States
50 Hampshire Street, 5th Floor, Cambridge, MA 02139, United States
The Boulevard, Langford Lane, Kidlington, Oxford OX5 1GB, United Kingdom

Library of Congress Cataloging-in-Publication Data
A catalog record for this book is available from the Library of Congress

British Library Cataloguing-in-Publication Data
A catalogue record for this book is available from the British Library

ISBN: 978-0-12-810521-4

For information on all Academic Press publications visit our website at
https://www.elsevier.com/books-and-journals

Working together
to grow libraries in
developing countries

www.elsevier.com • www.bookaid.org

Publisher: Nikki Levy
Acquisition Editor: Nancy Maragioglio
Editorial Project Manager: Billie Jean Fernandez
Production Project Manager: Nicky Carter
Designer: Matthew Limbert

Typeset by TNQ Books and Journals

CONTENTS

LIST OF CONTRIBUTORS

Emanuel O. Alamu
International Institute of Tropical Agriculture (IITA), Southern Africa Research and Administration Hub (SARAH) Campus, Lusaka, Zambia

Lydia M. Chabala
University of Zambia (UNZA), Lusaka, Zambia

Terence Chibwe
International Institute of Tropical Agriculture (IITA), Southern Africa Research and Administration Hub (SARAH) Campus, Lusaka, Zambia

Godfrey Chigeza
International Institute of Tropical Agriculture (IITA), Southern Africa Research and Administration Hub (SARAH) Campus, Lusaka, Zambia

David Chikoye
International Institute of Tropical Agriculture (IITA), Southern Africa Research and Administration Hub (SARAH) Campus, Lusaka, Zambia

Martin Chiona
Zambia Agricultural Research Institute (ZARI), Mansa, Zambia

Olivier Crespo
University of Cape Town, Cape Town, South Africa

Katrien Descheemaeker
Wageningen University, Wageningen, The Netherlands

Jim Ellis-Jones
Agriculture-4-Development, Silsoe, Bedfordshire, United Kingdom

Therese Gondwe
International Institute of Tropical Agriculture (IITA), Southern Africa Research and Administration Hub (SARAH) Campus, Lusaka, Zambia

Sabine Homann Kee-Tui
ICRISAT, Bulawayo, Zimbabwe

Obert Jiri
University of Zimbabwe, Harare, Zimbabwe

Mohamed J. Kayeke
Mikocheni Agricultural Research Institute (MARI), Dar es Salaam, Tanzania

Kokou Kintche
International Institute of Tropical Agriculture (IITA), Central Africa, Kinshasa, Democratic Republic of Congo

Elias Kuntashula
University of Zambia (UNZA), Lusaka, Zambia

Olipa N. Lungu
University of Zambia (UNZA), Lusaka, Zambia

Paramu L. Mafongoya
University of KwaZulu–Natal, Pietermaritzburg, South Africa

George Mahuku
International Institute of Tropical Agriculture (IITA), East Africa, Dar es Salaam, Tanzania

Patricia Masikati
ICRAF Zambia, Lusaka, Zambia

Ackson Mooya
International Institute of Tropical Agriculture (IITA), Southern Africa Research and Administration Hub (SARAH) Campus, Lusaka, Zambia

Chrispen Murungweni
Chinhoyi University of Technology, Chinhoyi, Zimbabwe

Nhamo Nhamo
International Institute of Tropical Agriculture (IITA), Southern Africa Research and Administration Hub (SARAH) Campus, Lusaka, Zambia

Pheneas Ntawuruhunga
International Institute of Tropical Agriculture (IITA), Southern Africa Research and Administration Hub (SARAH) Campus, Lusaka, Zambia

John O. Omondi
Ben Gurion University of the Negev, Beer Sheba, Israel

Kabir Peerbhay
University of KwaZulu–Natal, Pietermaritzburg, South Africa

Alexander Phiri
Lilongwe University of Agriculture & Natural Resources, Lilongwe, Malawi

Trinity Senda
Matopos Research Institute, Bulawayo, Zimbabwe

Gevious Sisito
Matopos Research Institute, Bulawayo, Zimbabwe

Obert Tada
Chinhoyi University of Technology, Chinhoyi, Zimbabwe

FOREWORD

Climate change debate has received more global and national-level attention than most environmental issues in the 21st century. Evidence of the impacts of climate change has sharpened the arguments and presented a sense of urgency in designing action plans for resource-constrained habits of sub-Saharan Africa. This has had a ripple effect and influence on the funding decisions by major development partners in Africa and leading to institutional rearrangements, including programming reprioritization in the Consultative Group for International Agricultural Research (CGIAR) programs.

However, regardless of these arguments, one thing is for certain: the impacts of climate variability will affect the majority of poor farm families across southern Africa regardless of their location-specific conditions. Therefore, now more than ever, robust and resilient adaptation and mitigating smart agricultural interventions are needed. Agricultural response will be shaped by, but not limited to, a suite of integrated modern technologies, prepared institutions, and informed people, in particular, entrepreneurial youth, equipped with knowledge, and capacity to tackle extreme events. The book *Smart Technologies for Sustainable Smallholder Agriculture: Upscaling in Developing Countries* takes a holistic view and approach in discussing biophysical (climate and soils) and socioeconomic (institutions, people particularly the youth, stakeholder involvement, policies, markets, and management skills) issues as they relate to and will influence the response to climate variability and change.

At the point of finalizing this publication, major milestones in breeding for climate variability, soil fertility management in relation to on-farm poverty, and awareness of the importance of youth development shape the discussion contributing to agricultural transformation and intensification. In southern Africa, other global issues such as immigration will also have a strong bearing on labor and skills development, whereas new models will take food systems forward.

The book was borne out of the constant requests from farmers demanding production options that take into account extremes of weather coupled with soil degradation and youth unemployment. Its contents also cater to a wider audience of researchers, students, and extension personnel who work hand in hand with smallholder farmers. The authors passionately shared their experiences and invaluable views to advance adaptation to, and mitigation of, climate change impacts in southern Africa.

Dr. Nteranya Sanginga
Director General
International Institute of Tropical Agriculture
August 15, 2016

PREFACE

Climate change is projected to have large-scale negative effects on the economic, social, and development spheres of the inhabitants of southern Africa. In this region, more than 70% of the inhabitants rely on agriculture for occupation, food, and incomes. In the advent of global change, smallholder farm families, in particular, face a huge risk of derailed livelihoods as a result of the effects of climate extremes on agriculture. There is a need to translate both climate adaptation and mitigation strategies into action plans, which can be utilized by the farmers. This transformation needs to lead to increased agricultural production, accesses to, and utilization of climate smart technologies. Public goods in the form of research for development technical products are key in advancing smallholder agriculture now and in the future. Efforts to reach out to such farmers are required to avert potential widespread disasters in southern Africa. The building blocks for technical interventions revolve around the management of both yield limiting (soil quality, water, radiation/sunshine hours) and yield reducing factors, e.g., weed and disease pressure; beneficial and collaborating faunal and floral symbionts. Such smart agricultural technologies need to be adequately described, simplified for disseminated to smallholder farmers.

A decade has passed since the most popular models and forecasts on climate/weather variability and extremes were documented and discussed at international level. In the last 5 years, discussions have forged ahead and international commitments discussed targeting greenhouse gas emissions from many sectors including agriculture. These emissions have a bearing on, but not exclusively, temperature and rainfall patterns, which directly influence biomass production, dynamics of faunal communities associated with agricultural systems, floral patterns, and weed species distribution. In southern Africa, agricultural production systems are already burdened with the occurrence of extremes of weather and climate. There is, therefore, an urgent need to scale up the effort of both developing and disseminating climate smart agricultural technologies in an integrated manner. The solutions of combating devastating effects of climate extremes addressing farmers' risk and increasing resilience, i.e., concentrating on efforts centered on both mitigation and adaptation. The target has to be the wide range of agricultural practitioners who are opinion leaders and can influence effective utilization, while simultaneously reaching out to the majority of

322 million of farm families in southern Africa. Technologies that address these challenges need to be climate smart, sustainable, and with a capacity to support agricultural intensification. Climate-smart agricultural technologies (CSAT), defined as agricultural practices that sustainably improve agricultural production and incomes, adapt and contribute to systems resilience and at the same time reduce or remove greenhouse gasses. For the CSAT to deliver, supportive good practices in production, handling and processing, ample policy environment, improved dissemination approaches to reach large numbers of people, and particular attention at young future have to be considered in future agricultural investments and commitments. Application of smart technologies resonates with the overarching goals of lifting the majority out of poverty and reclamation of degraded agricultural land, which multinational organizations are working on. Equally important is the link between this promotion of smart agricultural technologies and the recently adopted United Nations Development goals particularly goal nos. 1 and 2 on no poverty and zero hunger, no. 8 on decent work and economic growth, no. 13 on climate action, no. 15 on life on land, and no. 17 on partnerships for the goals.

The book *Smart Technologies for Sustainable Smallholder Agriculture: Upscaling in Developing Countries* is structured into three main sections. Section 1 defines the scope and gives a background to climate smart agricultural technologies; section 2 describes the application of technologies to mitigate and adapt to climate and weather extremes drawing examples from those technologies that are relevant to the subregion; and section 3 looks at the approaches to reach millions in the dissemination of CSAT. The chapters in this book contribute directly to the debate on the three principles of climate smart agriculture, i.e., improver production, adaptation and systems resilience, and mitigation and reduced emissions.

Nhamo Nhamo, David Chikoye, and Therese Gondwe
Lusaka, Zambia

INTRODUCTION

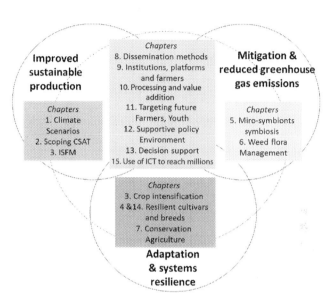

An illustration of chapters making up this book.

CHAPTER 1

Smart Agriculture: Scope, Relevance, and Important Milestones to Date

Nhamo Nhamo, David Chikoye
International Institute of Tropical Agriculture (IITA), Southern Africa Research and Administration Hub (SARAH) Campus, Lusaka, Zambia

Contents

1.1 INTRODUCTION

Agricultural production has stagnated in the past three decades due to a range of challenges farmers face in producing crops and livestock (Alexandratos and Bruinsma, 2012; Bajželj et al., 2014; Pandey, 2007; Steinfeld et al., 2006). Key among the challenges to smallholder agriculture are climate extremes and weather variability. These have exacerbated the extent to which the abiotic (e.g., soil degradation leading to infertile soils) and biotic (weeds, disease, and pests) constraints affect production (Balasubramanian et al., 2007; Nhamo et al., 2014; Sanchez, 2010). Climate change threatens the gains made in agriculture since the introduction of improved technologies (Funk and Brown, 2009; Schlenker and Lobell, 2010; Wheeler and von Braun, 2013). Urgent deployment of agriculture

Smart Technologies for Sustainable Smallholder Agriculture
ISBN 978-0-12-810521-4
http://dx.doi.org/10.1016/B978-0-12-810521-4.00001-3

1

technologies, which address the existing biotic and abiotic constraints, is required to harness the loss in agriculture production. Sanchez (2000) summarized the key global change scenarios relevant to developing countries and management of natural resource especially their direct effect on people, agriculture, carbon, water, nitrogen, and climate. The linkages between adaptation and mitigation to climate change leading to poverty reduction and improved natural resource management were elaborated. It is becoming clear that new investments and strategies for both underperforming and performing regions are required (Ray et al., 2012). Therefore, more focused research is needed to develop alternative options that will take agriculture production forward and provide food and nutrition to 9 billion people by year 2050. Furthermore, the scaling out of tried and tested technologies has to be priority for agricultural systems' transformation under changing climate.

1.2 SCOPING CLIMATE SMART AGRICULTURAL TECHNOLOGIES

Future agriculture will rely heavily on the application of modern technologies, which have the capacity of increasing the scale, efficiency, and effectiveness of production and delivery in all aspects of the commodity value chains. Sustainable agricultural production systems have gained favor from both producer and technology developers; public and private sectors worldwide. Climate and climate change will be a major consideration in the design, scaling up, and adoption/adaptation of agricultural technologies in the future (Alley et al., 2003). This is because extreme events have begun to take a toll on agricultural output against a background of increased food demands in the developing world. In southern Africa, major shift in the (1) rainfall pattern and (2) temperature incidences are increasingly becoming common (IPCC, 2007). Rainfall, depending on the latitudinal position, often start around October and end around May with a growing season length averaging between 3 and 8 months. More recently, the crop growing season has shrunk to barely 2–3 months (period between December and February) with the effect of reducing the potential production due to water unavailability for the greater part of the year (SADC YearBook, 2013). Similarly, heat waves have begun to affect evapotranspiration rates and hence agricultural production. There is no better timing for climate smart technologies that can alleviate the impending loss in agricultural production to avert hunger (famine), malnutrition, and ill-health.

Three Pillars of Climate Smart Agriculture

Climate smart agriculture, defined as agricultural practices that sustainably improve agricultural production and incomes, adapt and contribute to systems resilience, and at the same time reduce or remove greenhouse gasses (FAO, 2013), holds a promise to improve the agricultural productivity in Southern Africa (SA). A number of technologies have been developed to further the objectives of the three components of climate smart agriculture in Africa. These can be grouped differently depending on context and application relevancy. For the purposes of this book we have approached climate smart agriculture (CSA) as follows:

Pillar 1: To sustainably improve agricultural production and incomes encompasses a range of practices that are an important input in agricultural transformation, and crop intensification here is referred to as smart technologies. Smart technologies are therefore a basket of improved agricultural technologies and interventions, which have the following characteristics: (a) enhance rain-water productivity, conserve water, reduce surface evapotranspiration, and lead to increased water-use efficiency; (b) increase the capacity of production systems to withstand extreme temperatures (in both cold and hot weather) conditions, enable utilization of intrinsic temperature moderation at animal/plant cell level, and increase energy conservation within individual species and across systems; (c) increase capture and utilization of CO_2 for photosynthetic products. The majority of system agronomic practices fall in this category, e.g., integrated soil fertility management (ISFM), breeding for trait improvement, and conservation agriculture (CA).

Pillar 2: Mitigation and reduction of emissions include practices that utilize natural processes to minimize CH_4, CO_2, NO, and NO_2 production processes [greenhouse gases (GHGs) that when released into the atmosphere causes ozone layer depletion] and at the same time support pillar 1 in enhancing production (FAO, 2013; Stulina and Solodkiy, 2015; Vermeulen et al., 2012). Examples of technologies supportive of mitigation efforts include utilization of biological nitrogen fixation (BNF) and mycorrhizal associations to reduce reliance on mineral fertilizers.

Pillar 3: Adaptation and increased systems resilience include practices that address drivers of production at a higher level than technologies themselves. Some technologies described in pillar 1 contribute to immediate increase in yields but also contribute to systems' resilience, e.g., ISFM increases yields and improves soil structure through organic matter inputs. We consider the use of improved weed management techniques, increasing efficiency in the utilization of BNF, mycorrhizal associations, and nonafflatoxigenic fungal collaborations as ample examples for this pillar.

In addition, the definition of smart technologies goes further to include approaches for institutional improvements, organization of value chain players (innovation platforms for commodities), modernizing extension methodologies, supportive policies, and lobbying for engagement, increasing visibility of youth and women and putting technologies to use. Given the importance of technologies in transforming agricultural practices, smart technologies are the premise under which this book was developed. Fig. 1.1 illustrates the relationships among the three CSA pillars and the chapters of this book that address the same pillars.

1.2.1 Southern Africa Biophysical Characteristics

Southern Africa (SA) refers to a subregion of the African continent geographically covering 15 countries (Fig. 1.2) found between 10°N and 40°S (latitude) and 10°E and 50°E (longitude). Generally, the countries found in

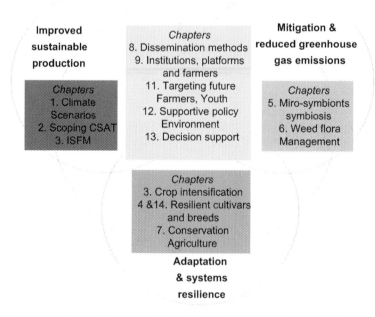

Figure 1.1 An illustration of the three pillars of climate smart agriculture and the chapters categorized under each pillar.

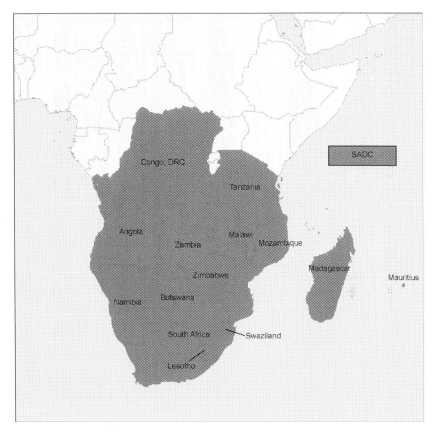

Figure 1.2 The map of the 15 countries found in the subregional grouping called Southern Africa Development Community (SADC). *Oosthuizen (2006).*

southern Africa include Angola, Botswana, Lesotho, Madagascar, Malawi, Mauritius, Mozambique, Namibia, Seychelles, South Africa, Swaziland, Zambia, and Zimbabwe. However, the Democratic Republic of Congo and Tanzania are members of the subregional grouping called Southern Africa Development Community (SADC) though they are geographically located in central and eastern Africa, respectively. The region has a combined population of 322 million and a land mass of 9.865 million km^2. The region is food self-sufficient but largely exports raw agricultural produce and imports processed food products. The region exports raw meat products to the European Union markets and most of the grain is traded within the region. Production occurs mainly in subhumid, semiarid, and arid areas with length of growing season ranging from 180 to 270 days, 70–180, and less than 70 days, respectively. The humid zone that covers a very small area has more than 270 days. The vegetation biome is a mixture of miombo woodlands (dominated by *Brachystegia speciforms* and *Brachystegia bohemi*), open savanna grasslands, and the dry savanna (dominated with *Acacia* tree species). Despite the large potential endemic in the subregion, a large import bill is paid by the regions on major commodities to include wheat, beef, soybean, and refined food products.

Agriculture in Southern Africa

Agriculture plays a pivotal role in the development and livelihoods of inhabitants of southern Africa (Sanchez, 2002). Initially, livestock rearing, i.e., large ruminants (cattle, donkeys, horses), small ruminants (goats, pigs, sheep), and poultry (chicken, ducks, turkeys, and ostriches) were the main occupation of most farmers but this has changed in the last five decades. Cultivation of crops has since taken over the sector supported by developed market linkages. Since the beginning of sedentary agriculture, cereals (O'farrell et al., 2009), (maize, rice, and wheat), root crops (cassava, sweet potato, yam), grain legumes (soybean, common bean, cowpea, chickpea, and fababeans), small grains (sorghum and millets), fruits (mango, apple, papaya, and guava) and leaf vegetables (cabbage, rape, broccoli, and spinach) have successfully been grown in SA (Rehm et al., 1984). Agriculture in SA depends on a unimodal rainfall pattern with the onset of the rainfall season around November and which ends around May. The pattern has of late been distorted heavily with much shorter seasons and longer mid-season drought experienced across the region presenting a huge challenge to farmer's families. Smart agricultural technologies are needed to overcome some of the climate variabilities that threaten and can cause massive crop and livestock losses. In particular rainfall patterns can result in water shortage for livestock, reduced biomass production for crops with high water footprint (Hoekstra and Chapagain, 2007), and increased pest and disease incidences for both (Rosenzweig et al., 2001).

Crop production in southern Africa has also evolved from the traditional rain-fed mixed crop livestock systems dominantly slash and burn (e.g., *Chitemene* and *fundikila* systems in Zambia) to modern systems dominated by mono-cropping systems, some of them irrigated; ample amounts of agrochemical are applied (e.g., Mathews et al., 1992). The level of development largely depends on the resource endowment of the farmers and the poor (who are in the majority about 60%) still rely on extensive low input agriculture systems partly slash and burn, whereas the well-endowed farmer (with access to resources and advanced technologies) operates larger cropping enterprises. On both extremes of the farmer typologies there are agricultural enterprises that are not sustainable, which contribute to land degradation (mainly through nutrient mining) and increased emission (from unbalanced nutrient management on farms); there is an opportunity to apply smart agriculture to rectify the situation.

Major soil groups found in southern Africa include ferrosols, acrisols, and nitisols. These are found in different environments, ranging from subhumid in highlands of Malawi, Mozambique, Zambia, and Zimbabwe to semiarid zones almost covering the whole region to arid zones in Namibia, Botswana, South Africa, and Zimbabwe and finally desserts in Botswana, Namibia, and Angola. Table 1.1 highlights some of the constraints associated with the major soil groups in southern Africa. With climate change there are fears that the arid zones will expand (Ngaira, 2007) northwards increasing the dessert margins in Botswana and Namibia (Kalahari and Namib desserts). Refining smart agriculture approaches surely will be handy for future food production.

1.2.2 Socioeconomic and Political Environment

Southern Africa has enjoyed political peace since the end of the Angola banditry activities about a decade back. The three countries to have experienced extended periods of internal civil conflict include Angola (upto the year 2003), Mozambique (until 1988), and Zimbabwe (ending in 1979). South Africa is the last country to get multiparty democracy in 1994. The stability has assisted all the 15 countries to maintain a constant economic growth rate [regional average (gross domestic product) GDP growth of about 4%]. The regional grouping SADC has also supervised the member countries into steps to prevent conflicts from recurring. Trade has improved and there are trade agreements aimed at reducing tax burdens of inhabitants involved in cross-border trading (SADC Yearbook, 2011).

Agriculture is an important sector for economies of SADC countries and there are policies that broadly support technology development and application across the region. However, challenges still exist in the area of biotechnology; in particular, there are no coherent policies dealing with genetically modified food and feed across the region. More work is required in developing policies supportive of agricultural technologies for smart agriculture to be applied successfully in southern Africa.

1.2.3 Recent Extreme Events Recorded in Southern Africa

Smart technologies have an important contribution to make in agricultural and livelihood development in southern Africa. In recent years,

Table 1.1 Characteristic of major soil types in southern Africa and limitations

Soil type	Main constraints	Degradation	Corrective measures
Alfisols	Salinity, carbonates dominate soil chemistry	Erosion, low OM, N and P, sodicity	Drainage, tillage, salt-tolerant crops
Andosol and nitisols	Fertile volcanic soils, P fixing, Mn toxicity, water and nutrient holding capacity limited	Erosion, acidity, low organic matter, low N and P	Use deep capture from subsoil, organic, and mineral fertilizers
Cambiols	Moderate nutrient content		
Ferrasols and acrisols	Al and Mn toxicity, high P fixation, low nutrient and water-holding capacity, low nutrient content and susceptible to erosion	Highly weathered soils, low organic matter	Manage nutrients to minimize leaching and P fixation
Ferric luvisols	Low nutrient content	Erosion, loss of nutrient N and P, soil acidity	N and P sources to reduces deficiencies
Fluvisols, Gleysols	Poor drainage		
Lithosols	Shallow and found in dry areas		
Vertisol	Heavy black soils, medium mineral reserves, narrow moisture availability range, erodibility, and flooding	Soil structure, reduced OC, high clay activity	Drainage, bunding, stone bunds and terraces on slopes

OC, organic carbon; OM, organic matter content

ecosystem level productivity has been challenged by extremes of weather and climate. Table 1.2 shows some of the stresses emanating from extreme events that affect agriculture production in southern Africa. In southern Africa skewed rainfall patterns have been experienced in recent times leading to floods along the Zambezi basin affecting livelihoods of producers in Malawi, Mozambique, and Zambia. Similar floods were experienced in the year 2000 resulting in widespread destruction of property mainly in Mozambique and along the Zambezi valley in general, i.e., areas in northern Zimbabwe, southern Malawi, and Zambia (Mirza, 2003). The extent to which technical interventions will remain relevant depends on their contribution to sustainability of production systems: capacity to service a heterogeneous farming community, service million hectares under crop production, livestock and wildlife areas, protect the two main watersheds (Limpopo and the Zambezi watersheds) hosting a number of power generating water reservoirs. Similar services will need to be extended with due consideration of millions of hectares under mixed crop-livestock farming systems.

Table 1.2 Extreme events recorded in southern Africa in the recent past and the effect on human life

Climate variable	Nature of variation	Resultant effect	Countries affected
Rainfall	Extreme rainfall activities	Flooding along the Zambezi basin	Malawi; Mozambique
	Delayed onset of rainfall season (2015–6 cropping season)	Late crop establishment	SADC except Tanzania and DR Congo, Angola
Temperature	Extreme hot weather (heat wave during last quarter of 2015)	High temperature during October and November	SADC except Tanzania and DR Congo, Angola
	Extreme cold weather (in the first quarter of 2016)	Disrupted crop growth	Mainly South Africa but also Malawi, Zambia, and Zimbabwe

DR, democratic republic

1.2.4 Supportive Initiatives in Agriculture Development in the Last 10 Years

African head of governments have realized the importance of agriculture and the role it can play in driving development, economic growth, and welfare of the population. To this end, several initiatives have been drafted by African governments to foster regional integration, trade, and common vision in developing the agriculture sector. The comprehensive African agriculture development program, launched in 2002, is an agreement by African governments aimed at poverty and hunger reduction through an agriculture-led economic growth. It is one of the major components and contribution of the New Partnership for Africa's Development since its inception. Equally important is the agenda 2063, a 50-year long plan for agriculture development in Africa, which was put in place in 2013. To strengthen these efforts, the Malabo declaration of 2015 was also adopted by African leaders. In this declaration, governments declared their desire for accelerated agriculture growth and transformation for shared prosperity and improved livelihoods. In 2015, the United Nations led the process of planning for the next generation of goals culminated in the adoption of 17 Sustainable Development Goals, guiding agriculture and expounding the desire for reduced poverty, unemployment, environmental degradation, hunger, and malnutrition. In 2016, the African Development Bank also added to the list a program called Feed Africa, which is pined on a transformation agenda in African Agriculture. The current discourse about private sector led agricultural transformation in Africa is protechnology development and application to a wider audience and probusiness in farming.

Among the key recommendation from the initiatives is the encouragement that each government should allocate about 10% of the GDP to the agriculture development. To date no country in southern Africa has followed this recommendation, on the contrary investment in research portfolios and support to extension services has actually gone down.

1.3 BUILDING IN SUSTAINABILITY WITHIN CLIMATE SMART TECHNOLOGIES

To avoid the impending crippling cycles of careless production practices that have led to an increase in the size of the ozone hole over Australia, future agricultural technologies need to be sustainable. We define sustainability as production systems that are intrinsically efficient and at the same time leave no negative footprint on the environment. Climate is one of the major environmental factors and therefore the smart technologies need to keep clean of the drivers of negative global climatic changes. Sustainability also entails efficiency to a level where producers are able to increase returns to land, labor, and machinery investments. However, farmers are heterogeneous and have no one size fit all solution, which will resolve their problems.

Sustainable agriculture has generally been defined as an integrated system of plant and animal production practices that is economically viable, socially supportive of consumers, and farm families and ecologically sound. Sustainable agricultural intensification is defined as producing more output from the same area of land while reducing the negative environmental impacts and at the same time increasing contributions to natural capital and the flow of environmental services. The three common approaches on sustainable intensification have been (1) increased yields per hectare, (2) increasing cropping intensity by increasing the number of crops grown on a piece of land (two or more crops), or (3) change of land use value from low-value crops to high-value commodities (Pretty et al., 2011). In recent years, sustainable intensification has taken center stage in terms of discussion. Adoption of intensification options by smallholder farmers will depend on a number of factors.

Hassan and Nhemachena (2008) identified access to markets, extension and credit services, technology and farm assets (labor, land, and capital), government policies, education, and information on adaptation as determinants of adaptation to climate change on smallholder farms. Using a criterion made up of 10 socioeconomic indicators, farmers were categorized into four groups to discuss options for intensification (Fig. 1.3). Fig. 1.3 shows some of the farmer categories and the level of investment and sophistication of operation.

Based on the four farmer categories, the most affected will be smallholder farms group 1, whereas smallholder group II and small-scale commercial group 1 represent farmer with capacity to adaption to climate change through input-led approach.

Table 1.3 shows some of the attributes that sustainable agriculture systems endeavor to achieve in the long run and share with climate smart agricultural technologies. The major difference between the two concepts, which definitely developed at different times, is the emphasis placed on some aspects of the definitions relative to the other.

In defining climate smart agriculture there is an apparent emphasis on adaptation and mitigation to the varied effects and negativities caused by the extreme global events recognition of the importance of the

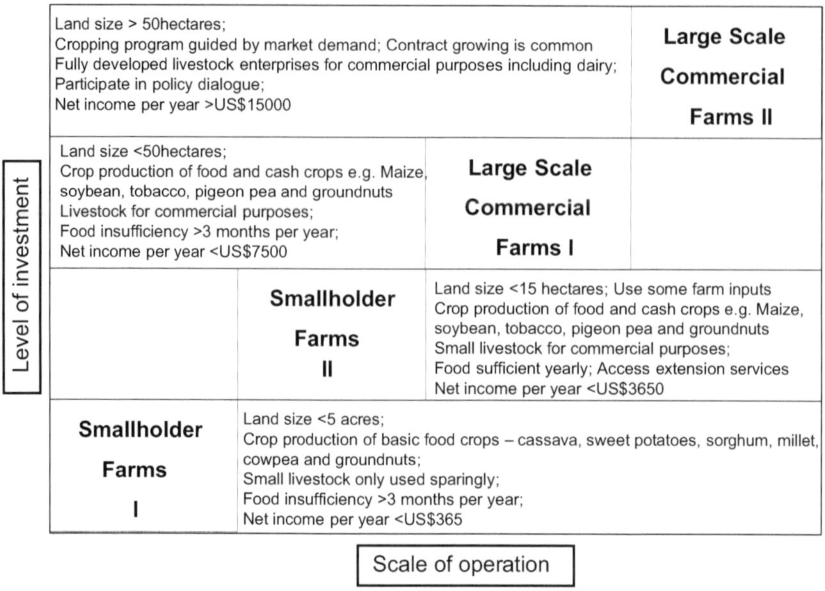

			Land size > 50hectares; Cropping program guided by market demand; Contract growing is common Fully developed livestock enterprises for commercial purposes including dairy; Participate in policy dialogue; Net income per year >US$15000	**Large Scale** **Commercial** **Farms II**
Level of investment		Land size <50hectares; Crop production of food and cash crops e.g. Maize, soybean, tobacco, pigeon pea and groundnuts Livestock for commercial purposes; Food insufficiency >3 months per year; Net income per year <US$7500	**Large Scale** **Commercial** **Farms I**	
		Smallholder **Farms** **II**	Land size <15 hectares; Use some farm inputs Crop production of food and cash crops e.g. Maize, soybean, tobacco, pigeon pea and groundnuts Small livestock for commercial purposes; Food sufficient yearly; Access extension services Net income per year <US$3650	
	Smallholder **Farms** **I**	Land size <5 acres; Crop production of basic food crops – cassava, sweet potatoes, sorghum, millet, cowpea and groundnuts; Small livestock only used sparingly; Food insufficiency >3 months per year; Net income per year <US$365		

Scale of operation

Figure 1.3 Farmer typologies that can be used in targeting both technologies and climate adaptation strategies for increased resilience.

Table 1.3 Commonalities between climate smart agriculture and sustainable agriculture

	Key attributes
Climate smart agriculture (FAO, 2013)	• Sustainably increase productivity and incomes • Adapting and building resilience to climate change • Reducing and/or removing greenhouse gas emissions
Sustainable intensification (Pretty et al., 2011)	• Utilizing crop varieties and livestock breeds with a high ratio of productivity to use of externally and internally derived inputs • Avoiding the unnecessary use of external inputs • Harnessing agroecological processes such as nutrient cycling, biological nitrogen fixation, allelopathy, predation, and parasitism • Minimizing the use of technologies or practices that have adverse impacts on the environment and human health • Making productive use of human capital in the form of knowledge and capacity to adapt and innovate and social capital to resolve common landscape scale problems • Quantifying and minimizing the impacts of system management on externalities such as greenhouse gas emissions, clean water availability, carbon sequestration, biodiversity, and dispersal of pests, pathogens, and weeds

phenomenon in the present time. It is our hypothesis that as systems evolve other urgent matters will arise; hence the term "smart agriculture", which is both forward looking and encompassing of systems. There are direct and indirect relationships among factors of production in agriculture and global change; hence the need for approaching the future agricultural systems using integrated technologies. Application of smart agriculture entails improved and efficient use of resources such as external inputs in production systems, e.g., energy. Increased use of fossil fuel on a highly mechanized farm could lead to an increase in greenhouse gas emissions and compaction of soils further compounding the problems of extreme weather events. Therefore, there is a need for moderation and determination of suitable threshold levels of application of improved technologies. With regards to energy use on the farm, the application of cleaner sources of energy (those with low environmental footprints) will take center stage in the future designs of agricultural systems. Systems will strive to increase the proportion of energy supply from both solar and wind sources, two well-documented sources of renewable energy.

The future of smart agriculture will also depend on the progress toward building soil health. Soil health, defined as the continued capacity of soil to function according to its potential and changes over time due to human use and management or to natural events (Doran and Safley, 1997), is an important attribute of productive soils. Maintaining soil health is only possible in situations where returns to investment (land, labor, and inputs) are enhanced, proper management of tillage (including use of appropriate implements), soil organic matter, fertility inputs (to avoid plant nutrient mining), and losses of excess nutrients (to reduce leaching and gaseous losses to the atmosphere) from such a system are of high relevance. Similarly, negative water footprints need to be minimized through the application of smart agriculture; crop cultivars and cropping systems and livestock technology systems with high water-use efficiencies are urgently required.

Soil health in southern Africa is highly associated with several paradigms and groups of technologies: best-bet technologies (since the 1980s), ISFM (since the 1990s), CA (since the years around 2000); the contribution of these are detailed in Chapter 3, 5, and 7. A lot of ground has been covered and more needs to be done in the future to refine extension messages coming out of the development of these. It is now apparent that advocating for blanket application of technologies is misplaced and can plunge farming communities into poverty due to poor returns to investment.

1.4 RELEVANCE OF SMART TECHNOLOGIES IN SOUTHERN AFRICA

Improved technologies were the backbone of the Green Revolution experienced in Asia in the 1970s. Three notable groups of technologies were among the leading technical interventions, which contributed to the huge increase in productivity, i.e., breeding for improved yield (crop stature and biotic stresses), widespread use of fertility inputs, and management of water on cropping lands. The prospects of a Green Revolution in sub-Saharan Africa are not going to depend only on the three groups mentioned above but more importantly on a wide scale adoption of technologies to solve the impending biotic stresses, e.g., weeds, pests, and disease, labor-saving technologies in mechanization, efficient use of bacterial and fungal microsymbionts in cropping systems to harness nitrogen from the atmosphere and nutrient absorption on the root cortex. Besides these, southern Africa also requires supportive institutional arrangements to enhance the interventions.

The context of southern Africa is that policies supportive of improved agricultural technologies are either present in general forms or absent completely and often the demand for such instruments is driven by a very small section of the population. This can be attributed to general low literacy levels in the majority of agricultural producers, alternatively the government-led processes of putting policies in place are tedious and only a small number has input to the whole process.

Given the global change, which is upon the region, relevant technologies need to be time and target bound for all purposes. Weather seems to be ever changing and designing technologies and applying them over extended periods of time has been a major drawback in some instances. The most quoted example is that of blanket fertilizer recommendations developed for the majority of crops in the 1960s and 1970s. Although the recommendations were supported very well by extension systems, it became apparent that it is incorrect to assume uniform application across evidently heterogeneous environments in a number of agroecological zones. The practice of applying 200 kg compound D and 200 kg (ammonium nitrate) AN or Urea indiscriminately across all fields of different soil types and topographical locations has tremendous environmental effects in the long run when followed precisely. Similarly, farmers use cultivars that were developed and released for use decades ago, with the assumption that the breeding products will resolve current challenges. Such practices, where inappropriate technologies are applied by farmers, may have well contributed to the observed stagnation of yield in the past two decades.

Smart technologies have a big role to play in agricultural transformation for Africa to be able to feed its population and also export excess value-added products. Production based on expansion of land under cropping cannot continue as land is limited and is nonrenewable and, agriculture and other landuses contribute about 24% of the global greenhouse gas emissions. Appropriately designed sustainable intensification defined here as producing more output from the same area of land while reducing the negative environmental impacts and at the same time increasing contributions to natural capital and the flow of environmental services (Pretty et al., 2011). Smallholder farmers in southern Africa are among the most heterogeneous group varying from land-use sizes to resource endowment and investment objectives for each unit. Fig. 1.3 shows the linkages between farmer's investment objectives and the potential scale of operation and adoption of technologies.

1.5 THE ECONOMICS OF APPLYING SMART TECHNOLOGIES IN AGRICULTURE

Agriculture is the main source of income and employment for at least 60% of the population in southern Africa. If the millennium development goals of reducing poverty and hunger are to be achieved, agricultural systems need to be financially rewarding to those engaged in it. Furthermore, for sustainable intensification to reap financial benefit, basic economic principles should apply to the enterprises: economies of scale on large operations, mechanization to increase efficiency, and sustained demand for the produce (demand > supply). There is a need for research to develop and provide threshold figure so that farm units design operations based on known targets and cash flow projections. In southern Africa market information and price guarantees of produce only apply to a few commodities, a situation which needs urgent redress. In addition, technologies need to lift farmers out of poverty and reduce risk on investment.

Agriculture as a business is prone to numerous uncertainties in the form of systemic and covariant risk. To date, only a few index-based insurance products are well developed for the sector (Rao, 2010). Millions of farmers who base their livelihood on agricultural sector run uninsured enterprises. In the advent of climate change, farmers with significant uninsured investment on the farms, will face higher risk levels (Barnett and Mahul, 2007; Chantarat et al., 2013). Hence there is a need for an increased role of enterprise insurance to reduce risks of loss of investment resulting from natural and other disasters, e.g., drought or floods. However, in smallholder agriculture, the fragmentation of farms found in remote areas weakens the support for insurances as these may not be financially viable and hence carry limited attraction to insurers. In other words, the major challenge though is to get models, which will incentivize the insurance providers. First, if the insurer stands to lose money when a drought affects the whole region (as is often the case) and payouts need to be made to all the subscribers then the viability is questionable. Second, the assessment of risk and losses in agriculture enterprises are data intensive making assessments extremely difficult. Despite these challenges, resolving insurance issue in agriculture will most probably increase interest and stabilize investment in agriculture in the future.

Fragmented farms are a characteristic feature of smallholder agriculture in southern Africa and often explain the variation in management practices across the small pieces of land. There are advantages and disadvantages of fragmented farms and it is not clear which side outweighs the other across the region. The costs of fragmentation include increased traveling time between fields, higher transport costs, reduced scope of mechanization, lower labor productivity, negative externalities, loss of land for boundaries, reduced investment opportunities, and greater potential for disputes (Blarel et al., 1992). On the other hand fragmentation can allow farmers to manage risk better, overcome seasonal labor bottlenecks, and provide a better matching between soil types and necessary food crops. Diagnostic farm surveys have shown that land holding sizes vary averaging in some case to less than 0.3 ha per household. Table 1.4 shows some gross margin calculation on incomes from commonly grown crops. It is not financially possible on such small portions of land (e.g., < 1ha) to generate agricultural produce that will lift a family of six (average household size in southern Africa) out of poverty and sustain reasonable livelihoods. Production figures and prices in Table 1.4 confirm this hypothesis, a position technocrats need to discuss with farmers. Although bulking centers have worked well, e.g., for cotton maize and groundnuts, to provide closer markets and aggregate product from small farmers, it is highly unlikely that innovations and smart agriculture will contribute significantly to the profitability of enterprise of farmers with small land holdings. Fragmentation of farms will affect total production,

Table 1.4 Production characteristics for major crops produced under smallholder conditions on 1 ha basis in southern Africa

Crop	Area under production (ha)	Average yield (kg/ ha)	Selling price/ ton ($)	Production cost/ha ($)	Gross margin ($)
Maize	13,847,442	604 (1620)	265	529	−369
Groundnuts	2,030,180	659	872	241	334
Cassava	2,567,278	5300 (16271)	284	174	278
Sorghum	1,704,802	882	219 (352)	174	19
Soybean	362,049	1393	594	467	360
Rice	2,132,569	860 (1890)	644	467	87

SADC Yearbook (2013).

level of investment, and efficiency of resource use; therefore research should give guidance on farm sizes that can benefit from aggregated operations and marketing logistic for profitability.

1.6 INVESTING INTO TARGETED TECHNOLOGIES FOR THE FUTURE

Climate smart agricultural technology applications range in size of investment from an animal-drawn feterrali riper-plantar useful at smallholder farms to a highly mechanized tillage and fertilizer application machinery used in "precision agriculture" or a multimillion dollar fertilization plant often suitable for large production plants, in the scale of applications from one acre farm operations to thousands of hectares of mono-cropping clean cultivation, external input-intensive cash crops on large-scale farms to low-input traditional food and marginalized minor crops on smallholder remote farms, all meant to respond to extremes of temperature and skewed rainfall patterns and reduce emission. Regardless of the scale, climate smart technologies will require organized investment plans for sustainable and ever developing agricultural production practices. The guiding principle to the investment in all the technologies needs to be the capacity of the farming system to produce without compromising the future potential. In this regard where intensification has been adopted, the increasing benefits from agro-chemical and machinery (compared to manual systems of labor) need to take soil health and sustainability issues into consideration.

The greatest yield mover of yester-years was the breeding for higher yield programs as they have contributed to about 90% of the observed yield gains in most cultivated crops in general. Similar trends were observed in livestock from large to small mammals and birds. Massive investment has gone into improving breeding techniques, however, the future yield gains may not come so much from breeding but rather from associated improved crop/animal management practices, including good agronomic practices, resolution of pests and disease challenges. Similarly mechanization in post-harvest handling and processing, market information systems, and policies are key investment areas required in smallholder agriculture.

To bring about the desired large-scale impact we hypothesize that a holistic approach to investments targeting need to consider already available infrastructure and ride on it. This can be operationalized by targeting specific agroecological zones, certain commodity value chains, and development zones where the cost of doing transactions is lowest. This is because it has been observed that trying to improve productivity in isolated and fragmented areas with no access to both input and output markets is an uphill struggle.

In addition, there is an urgent need for a system's approach where a suit of technologies are disseminated as a basket of options for farmers. Often farmers adopt technologies in a step-by-step fashion and this has to be taken into consideration. This book takes a similar approach by combining a range of technologies and at the same time addresses institutional and policy issues (Fig. 1.2).

1.7 CONCLUSION

In conclusion, agricultural technology generation and development practitioners have a challenging task of proving strategic initiative to feed a growing population by doubling production in the next decade or two. Suitable smart agricultural technologies target large-scale producers to double their efforts by increasing returns to investment and land, assist smallholder farms to achieve household food and income security, and at the same time harness the spiraling greenhouse gas emissions from agriculture. To this end several milestones have been achieved within the context of individual technological development: increased availability of improved varieties and species of both crops and animals using improved breeding techniques, availability of good agronomic practices, improved mineral fertilizer use albeit its high prices due to importations and energy costs. Information is available for scaling out some technologies; however, gaps exist in the development of smart agriculture. More can be achieved through climate change responsive technology development and regiment coupled with efforts to integrate a suit of these technologies to increase the resultant synergies.

Decision guides to assist farmers adopt efficient resource use, application of site-specific production practices, marketing logistics are urgently needed to resolve the challenge farmers have in targeting yields from available inputs and applying agronomic interventions in a timely manner.

REFERENCES

Alexandratos, N., Bruinsma, J., 2012. World Agriculture Towards 2030/2050: The 2012 Revision. (ESA working paper no. 12-03).

Alley, R.B., Marotzke, J., Nordhaus, W.D., Overpeck, J.T., Peteet, D.M., Pielke, R.A., Pierrehumbert, R.T., Rhines, P.B., Stocker, T.F., Talley, L.D., Wallace, J.M., 2003. Abrupt climate change. Science 299 (5615), 2005–2010.

Bajželj, B., Richards, K.S., Allwood, J.M., Smith, P., Dennis, J.S., Curmi, E., Gilligan, C.A., 2014. Importance of food-demand management for climate mitigation. Nature Climate Change 4 (10), 924–929.

Balasubramanian, V., Sie, M., Hijmans, R.J., Otsuka, K., 2007. Increasing rice production in sub-Saharan Africa: challenges and opportunities. Advances in Agronomy 94, 55–133.

Barnett, B.J., Mahul, O., 2007. Weather index insurance for agriculture and rural areas in lower-income countries. American Journal of Agricultural Economics 89 (5), 1241–1247.

Blarel, B., Hazell, P., Place, F., Quiggin, J., 1992. The economics of farm fragmentation: evidence from Ghana and Rwanda. The World Bank Economic Review 6 (2), 233–254.

Chantarat, S., Mude, A.G., Barrett, C.B., Carter, M.R., 2013. Designing index-based livestock insurance for managing asset risk in Northern Kenya. Journal of Risk and Insurance 80 (1), 205–237.

Doran, J.W., Safley, M., 1997. Defining and assessing soil health and sustainable productivity. In: Pankhurst, C., et al. (Ed.), Biological Indicators of Soil Health. CAB International, Wallingford, UK, pp. 1–28.

FAO, 2013. Climate-Smart Agriculture Source Book. Food and Agriculture Organization of the United Nations, Rome, p. 545.

Funk, C.C., Brown, M.E., 2009. Declining global per capita agricultural production and warming oceans threaten food security. Food Security 1 (3), 271–289.

Hassan, R., Nhemachena, C., 2008. Determinants of African farmers' strategies for adapting to climate change: multinomial choice analysis. African Journal of Agricultural and Resource Economics 2 (1), 83–104.

Hoekstra, A.Y., Chapagain, A.K., 2007. Water footprints of nations: water use by people as a function of their consumption pattern. Water Resources Management 21 (1), 35–48.

IPCC, 2007. Climate Change: Impacts, Adaptation and Vulnerability. In: Parry M.L., Canziani O.F., Palutikof J.P., van der Linden P.J., Hanson C.E. (Eds.), Cambridge University Press.

Mathews, R.B., Holden, S.T., Volk, J., Lengu, S., 1992. The potential of alley-cropping in improvement of cultivation systems in the high rainfall areas of Zambia I. Chitemene and Fundikila. Agroforestry Systems 17 (3), 219–240.

Mirza, M.M.Q, 2003. Climate change and extreme weather events: can developing countries adapt?. Climate policy 3 (3), 233–248.

Ngaira, J.K.W., 2007. Impact of climate change on agriculture in Africa by 2030. Scientific Research and Essays 2 (7), 238–243.

Nhamo, N., Rodenburg, J., Zenna, N., Makombe, G., Luzi-Kihupi, A., 2014. Narrowing the rice yield gap in East and Southern Africa: using and adapting existing technologies. Agricultural Systems 131, 45–55.

O'farrell, P.J., Anderson, P.M.L., Milton, S.J., Dean, W.R.J., 2009. Human response and adaptation to drought in the arid zone: lessons from Southern Africa. South African Journal of Science 105 (1–2), 34–39.

Oosthuizen, G.H., 2006. The Southern African Development Community: the organisation, its policies and prospects, Institute for Global Dialogue.

Pandey, S.K., 2007. Approaches for breaching yield stagnation in potato. Potato Journal 34 (1–2).

Pretty, J., Toulmin, C., Williams, S., 2011. Sustainable intensification in African agriculture. International Journal of Agricultural Sustainability 9 (1), 5–24.

Rao, K.N., 2010. Index based crop insurance. Agriculture and Agricultural Science Procedia 1, 193–203.

Ray, D.K., Ramankutty, N., Mueller, N.D., West, P.C., Foley, J.A., 2012. Recent patterns of crop yield growth and stagnation. Nature Communications 3, 1293.

Rehm, S.E., Rehm, S.V., EA Espig, G., Rehm, S.Z., 1984. The Cultivated Plants of the Tropics and Subtropics: Cultivation, Economic Value, Utilization. Margraf (No. 581.60913 R345).

Rosenzweig, C., Iglesias, A., Yang, X.B., Epstein, P.R., Chivian, E., 2001. Climate change and extreme weather events; implications for food production, plant diseases, and pests. Global Change & Human Health 2 (2), 90–104.

SADC Statistics Yearbook, 2011. SADC Protocols and Other Legal Key Instruments. SADC, Gaborone. http://www.sadc.int/information-services/sadc-statistics/sadc-statiyearbook/.

SADC YearBook, 2013. Southern Africa Development Community (SADC) Statistical Year Book. http://www.sadc.int/information-services/sadc-statistics/sadc-statistics-yearbook-201/.

Sanchez, P.A., 2000. Linking climate change research with food security and poverty reduction in the tropics. Agriculture Ecosystems and Environment 82, 371–382.

Sanchez, P.A., 2002. Soil fertility and hunger in Africa. Science 295 (5562), 2019.

Sanchez, P.A., 2010. Tripling crop yields in tropical Africa. Nature Geoscience 3 (5), 299–300.

Schlenker, W., Lobell, D.B., 2010. Robust negative impacts of climate change on African agriculture. Environmental Research Letters 5 (1), 014010.

Steinfeld, H., Wassenaar, T., Jutzi, S., 2006. Livestock production systems in developing countries: status, drivers, trends. Revue Scientifique et Technique (International Office of Epizootics) 25 (2), 505–516.

Stulina, G., Solodkiy, G., 2015. The effect of climate change on land and water use. Agricultural Sciences 6, 834–847.

Vermeulen, S.J., Campbell, B.M., Ingram, J.S.I., 2012. Climate change and food systems. Annual Review of Environment and Resources 37, 195–222.

Wheeler, T., von Braun, J., 2013. Climate change impacts on global food security. Science 341 (6145), 508–513.

CHAPTER 2

Climate Scenarios in Relation to Agricultural Patterns of Major Crops in Southern Africa

Paramu L. Mafongoya[1], Kabir Peerbhay[1], Obert Jiri[2], Nhamo Nhamo[3]

[1]University of KwaZulu–Natal, Pietermaritzburg, South Africa; [2]University of Zimbabwe, Harare, Zimbabwe; [3]International Institute of Tropical Agriculture (IITA), Southern Africa Research and Administration Hub (SARAH) Campus, Lusaka, Zambia

Contents

2.1 INTRODUCTION

Agriculture in southern Africa plays a critical role in the formal and informal economy (Chen, 2005). Agriculture is critical in sustaining rural livelihoods and food security (Godfray et al., 2010; Vermeulen et al., 2012; Lobell et al., 2008). Agriculture is directly dependent on climatic factors such as solar radiation, temperature, and precipitation. Therefore climatic variability dictates the potential of crop and livestock production systems in general

Smart Technologies for Sustainable Smallholder Agriculture
ISBN 978-0-12-810521-4
http://dx.doi.org/10.1016/B978-0-12-810521-4.00002-5

21

and more specifically the regions or locations suitability, the characteristics of cultivars and breeds to be used, range of options from which choices can be made, and the cropping calendars to follow (Tingem and Rivington, 2009; Unganai and Kogan, 1998). The agricultural sector is inherently sensitive to shifts in climate; hence the need to manage some of the uncertainties. Changes to the climatic variables may alter agricultural productivity in general, crop yield, and livestock potential leading to altered food production pattern in various ways (Davis, 2011; Rosenzweig et al., 2001). Table 2.1 summarizes the impacts of these climatic shifts to agricultural production in southern Africa.

Climate change is the long-term shift and fluctuation in the statistics of weather patterns of an area as reflected in weather parameters such as precipitation, temperature, humidity, wind, and seasons (IPCC, 2007; Ngaira, 2007). The natural variability has been found to play a subsidiary role in recent climate changes compared to the anthropogenic increase in GHG (Crowley, 2000). All these climatic conditions are likely to pose new challenges for various farming systems, crops, and regions (Fischer et al., 2005). With 2°C increase in temperature and 10% reduction in rainfall, the maize yield in southern Africa is expected to experience a reduction of 0.5 t/ha (Schulze, 2007). The greatest impact of production is expected to be in most marginal areas where rainfall is low and irregular irrigation is practiced. Smallholder agricultural systems are at high risk due to high dependence on rainfed agriculture as well as fewer capital resources and management technologies available to them. It is critically important to identify the full set of drivers changing agricultural productivity that tends to be complex and integrated at various levels.

Despite this, a decline in crop production as a result of climate change is likely to adversely affect food security because it will increase food prices. Southern Africa is likely to be impacted by future climate change as the latest climate change projections for the region indicate that both temperature and evapotranspiration are likely to increase in the 21st century. Climate change is likely to alter the magnitude, timing, and distribution of storms, which produce flood events as well as frequency and intensity of drought events (Fauchereau et al., 2003).

Southern Africa has existing critical vulnerability that exacerbates the effects of future climate change in most sectors due to their dependence on the natural environment and resources for livelihoods. Recurring hazards lead to both direct and indirect and secondary impacts on society and agricultural systems. Direct losses are defined here as the actual financial value

Table 2.1 Direct and indirect impacts of projected climate change on crop and livestock production and the socioeconomic implications to communities living in southern Africa

Production system	Nature of impact	Description of effect	References
Crop production	Direct impacts	• Even small increases in mean temperature of between 1 and 2°C are projected to lead to a decrease in crop productivity • Changes in temperature regimes could affect growing locations, the length of the growing season, crop yields, planting and harvest dates • Increased need for irrigation in a region where existing water supply and quality is already negatively affected by other stressors • Ecological hazards of droughts desertification and soil erosion may worsen making the areas unsuitable for crop production	Davis (2011), Mirza (2003), Ngaira (2007), Rosenzweig et al. (2001) and Unganayi and Kogan (1998)
	Indirect impacts	• Predicted higher temperatures are likely to negatively impact organic matter, thereby reducing soil nutrients • Higher temperatures may favor the spread of significant weeds, pests, and pathogens to a range of agricultural systems	
Livestock production	Direct impacts	• Changes in forage quality and quantity (including the availability of fodder crops) • Changes in water quality and quantity • Reduction in livestock productivity by increasingly exceeding the temperature thresholds above the thermal comfort zone of livestock, which could lead to behavioral and metabolic changes (including altering growth rate, reproduction, and ultimately mortality) • Increased prevalence of "new animal diseases" • Increases in temperature during the winter months could reduce the cold stress experienced by livestock, and warmer weather could reduce the energy requirements of feeding and the housing of animals in heated facilities	McMichael et al. (2007), Pelletier and Tyedmers (2010) and Thornton (2010)
	Indirect impacts	• Increased frequency of disturbances, such as wild fires • Changes in biodiversity and vegetation structure	
Socioeconomic/livelihood impacts		• Changes in incomes derived from crop and livestock production • Shifts in land use (including consequences of land reform) • Overall changes in food production and security	Ngaira (2007), Fischer et al. (2005)

of damage to and loss of capital assets, whereas secondary includes disruption of development plans, increased balance of payment deficits, increased public sector deficits and debt, and worsened poverty levels. Secondary impacts relate to the short- and long-term impacts on aggregate economic performance (Benson and Clay, 2000).

Understanding these climatic changes and their possible impacts on society are thus essential in critical sectors in southern Africa to improve strategic adaptation responses.

2.2 SOUTHERN AFRICA IN CLIMATE CHANGE AND HISTORICAL CHANGES

Four broad climatic zones have been used to describe Africa based on a combination of precipitation, temperature, and evapotranspiration, i.e., (1) arid and semiarid (Sahel region, Kalahari and Namib deserts), (2) tropical savanna grasslands (Sahel, Central, Southern Africa), (3) equatorial (The Congo region and the East African highlands), and (4) temperate (The Southeastern tip of South Africa) (Ngaira, 2007). Southern Africa is predominantly a semiarid region with high rainfall variability. The region is characterized by frequent droughts and floods. Work of Unganai and Kogan (1998) has shown recurrence of drought in the region during the 1982–4, 1991–2, and 1994–5 season, which led to massive crop losses. The region is also widely recognized as one of the most vulnerable regions to climate change because of low levels of adaptive capacity especially in rural communities combined with a high dependence on rainfed agriculture (IPCC, 2007). The use of weather forecasting in agricultural planning has remained low albeit its critical role. Furthermore, the current trend of investment in southern Africa has focused more on recovery from droughts and other disasters at the expense of developing the adaptive capacity of communities with high risk (Mirza, 2003).

The climate of southern Africa is strongly determined by the Pacific air circulation of the subcontinent in relation to the major circulation patterns of the southern hemisphere, complex regional topography, and the surrounding ocean movements. The resounding effect of these air mass circulations has explained in part some of the precipitation and temperature patterns observed in the region.

2.2.1 Rainfall

The region is described as predominantly semiarid with high seasonal and interannual rainfall variability. Extreme events such as droughts and floods occur frequently. The amount and seasonal distribution of rainfall are the most critical factors to consider when looking at rainfall across the region. There is a high degree of spatial variation in rainfall across the region. The average rainfall in the region is less than 1000 mm per year based on the period 1901–2009. Rainfall tends to decrease to the northeast and to the southwest (Fig. 2.1). Highest rainfall occurs in the highlands of eastern Madagascar, which receive 3100 mm per year. The majority of the region receives between 500 and 1500 mm per year, with more semiarid regions of the south receiving between 250 and 500 mm per year (Fig. 2.1). Rainfall over most of southern Africa is moderately seasonal. The majority of the rainfall occurs in the summer months of the year between October and March.

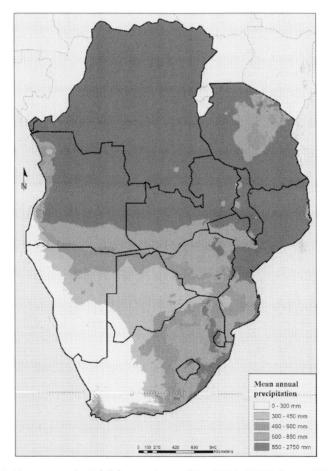

Figure 2.1 Mean annual rainfall for southern Africa (1901–2009). *Source: WorldClim— World Climate Data.*

Southern African interannual rainfall variability is known to be linked to the El Nino Southern Oscillation (ENSO) phenomenon. During the warm ENSO events dry conditions generally occur over much summer rainfall regions of southern Africa. In 1982/1983 below average rainfall and droughts in many parts of the region coincided with serious El Nino event. The influence of El Nino is strongest in the southeastern parts of the region and reaches a maximum in the last summer from January to March (2016). Other important determinants of rainfall patterns in southern Africa include the Inter-Tropical Convergence Zone (ITCZ). The ITCZ is a region characterized by high convergence activities resulting in high rainfall in several countries in summer months when its position shifts to the southern hemisphere.

2.2.2 Temperature

Southern Africa has a warm climate and much of the region experiences average warm temperatures above 17°C. Across the region mean annual minimum temperatures range from 3 to 25°C and mean annual maximum temperatures range from 15 to 36°C (Fig. 2.2). Frost is common in winter months on the central plateau and high altitude areas such as in South Africa and Zimbabwe. The greatest decrease in temperature is observed over the central plateau regions and the highland areas where the range difference can be 19°C. The coldest temperatures are experienced over South Africa, extending to southern parts of Namibia where temperatures average less than 15°C. Across the region average temperatures are mostly determined by a combination of latitude, elevation, and proximity relative to the coastlines of Indian and Atlantic Oceans (Chevalier and Chase, 2016).

2.3 CLIMATE CHANGE TRENDS IN SOUTHERN AFRICA

One of the best ways to understand how southern Africa regional climate may change in the future is to examine how it has changed in the past. This can be done by examining the observational records in southern Africa for evidence of climate trends over the last century.

2.3.1 Long-Term Observations

Climate observations are models used to determine changes in climate. However, most rural communities in southern Africa have always relied on indigenous knowledge systems to help them deal with climate variability

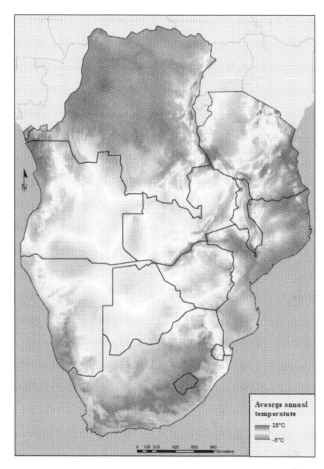

Figure 2.2 Mean annual temperature for southern African (1901–2009). *Source: WorldClim—World Climate Data.*

and change (Jiri et al., 2015). A key question when dealing with indigenous knowledge is how it can be integrated successfully with scientific knowledge to develop climate mitigation and adaptation strategies.

2.3.2 Temperature

There is strong evidence based on analysis of mean and maximum temperature trends that the region is getting warmer. These trends are shown in Fig. 2.3. From the 1970s these anomalies were almost persistent, approximately 0.8°C above the 1901–90 averages. The anomalies are also larger in more recent years, suggesting that the rate of increase in minimum and

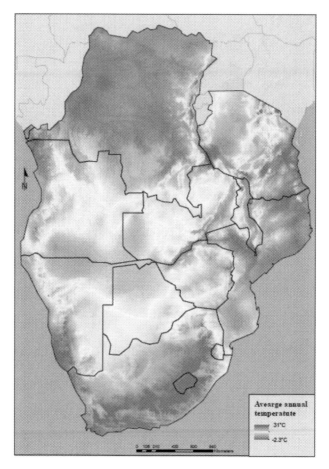

Figure 2.3 Projected mean annual temperature for southern Africa (2050). *Source: WorldClim—RPC 8.5, 2050.*

maximum temperatures is increasing (Jiri et al., 2015). These observations are consistent with detected increases in global annual air surface temperatures over southern Africa since 1900s (Kruger and Shongwe, 2004).

Trend analysis of temperatures across southern Africa reveals that annual minimum and maximum temperatures have increased at an average rate of 0.057°C per decade and 0.046°C per decade, respectively, between 1901 and 2009. Further analysis reveals that the periods of most rapid warming occur post-1990, a period for which the rate of increase for both average annual minimum and maximum temperatures is statistically significant.

After 1976, minimum temperatures began increasing by 0.27°C per decade and maximum temperatures by 0.25°C per decade (Davis, 2011). The larger rate of increase in minimum temperatures has been observed before (Alexander et al., 2006). Studies showed a general trend toward less severe cold events. After 1995, the highest observed maximum temperatures began to increase at a rate of 0.85°C, suggesting that the frequency of hot years is increasing.

2.3.3 Rainfall

Changes in rainfall are cryptic and difficult to detect. This is due to the fact that rainfall varies from place to place and from year to year across southern Africa. Existing evidence suggests no decrease in annual rainfall over southern Africa, however, certain characteristic of precipitation varied significantly across locations (Kruger, 2006). Other evidence has shown that interannual amount variability over southern Africa has increased since the late 1960s and droughts have become more intense and widespread in the region (Fauchereau et al., 2003). The pattern of anomalies shows that year-to-year rainfall variability is high across the region and has been a persistent feature of the region's climate for many years. The alternating patterns of above-normal and below-normal rainfall scenarios clearly illustrate that rainfall cycles fluctuate in the region. Extreme wet and dry years have been recorded, which resulted in floods and droughts. In 1999–2000 tropical cyclone Eline caused widespread flooding in southern and central Mozambique, southeastern Zimbabwe, and parts of South Africa and Botswana. In 1982–3 (El Nino year), 1986–7, and 1991–92 serious droughts were experienced that caused a decrease in crop and livestock production in many parts of the region (Vogel, 1994). The ENSO cycle is highly likely to be the main force responsible for many of the teleconnections in southern Africa rainfall patterns. This phenomenon though evident in southern Africa is absent or insignificant in other regions. In instances where records are available for a long time, there have been detectable increases in the number of heavy rainfall events. Studies in the regions have shown that the length of dry seasons and rainfall intensity have increased.

Over southern Africa there is evidence to suggest that temperatures have been increasing over the last century and that the rate of warming has been increasing, most notably in the last two decades. No clear evidence exists for changes in mean rainfall. The rainfall time series remain dominated by patterns of interannual rainfall variability.

2.4 FUTURE CLIMATE SCENARIOS OVER SOUTHERN AFRICA

A scenario is an incoherent, internally consistent, and plausible description of a possible future state of the world. It is not a forecast, rather scenario is an alternative of how the future can unfold (Davis, 2011). A set of scenarios is often adopted to reflect the possible range of future conditions, which can be based on changes in the climate system, socioeconomic circumstance, or other future changes (Davis, 2011).

2.5 DETERMINING FUTURE CLIMATE SCENARIOS

The Global Circulation Models (GCM) are the fundamental tools used for assessing the causes of past changes and inferring future changes. These are complex models based on the laws of physics, which represent interaction between the different components of the climate system such as the land surface, the atmosphere, and the oceans. Because future levels of greenhouse gas emissions in the atmosphere are dependent on the behavior or policy changes, whether or not we continue to depend on fossil fuels or switch to renewable energy sources, the models simulate climatic changes under a range of emission scenarios. Each scenario represents a plausible future. That there is a range of future possibilities which is an important concept in understanding clearly as it means that we can only suggest futures that maybe more likely than others.

The spatial resolution of GCM is too low to accurately represent the circulation patterns that determine climate at regional and local scales. To generate more detailed simulations at regional and local climate, two main types of downscaling methodologies maybe employed. These are statistical (empirical) and dynamical downscaling methods. Downscaling is the term used to describe the process through which the projections of change from GCMs are translated to the regional and local scales. The downscaled scenarios at a finer spatial scale are more useful for assessing local and regional impacts, adaptation, and developing policies.

2.5.1 Rainfall

Most scenarios generated by GCMs suggest the delay in the main rainfall season for large parts of southern Africa and increase in incidences of droughts over central southern Africa. Majority of the models are simulating a decrease in rainfall over most of the region (Fig. 2.4). There is also

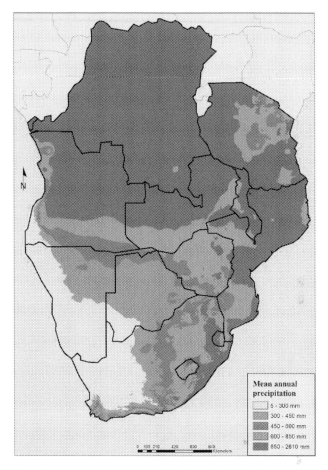

Figure 2.4 Projected mean annual precipitation by 2050. *Source: WorldClim—RPC 8.5, 2050.*

consistent simulated decrease in rainfall during summer across much of southern Africa. This is the period encompassing the start of the rains and suggests a reduction in early season rainfall.

2.5.2 Temperature

The models showed an increase in temperature under various scenarios of (greenhouse gas(es)) GHG emissions. Temperatures are expected to rise between 1 and 3°C. The weather of southern Africa is warming about twice the rate of global average (Scholes et al., 2015).

2.6 PROJECTED CHANGES IN EXTREME WEATHER EVENTS OVER SOUTHERN AFRICA

Climate change may manifest itself not only through changes in the long-term mean rainfall, temperature, and circulation patterns, but also through the increase in the frequency and intensity of extreme weather events. The concentration of GHGs in the atmosphere has enhanced the ability of the atmosphere to absorb and hold moisture. The excess heat and moisture can lead to increases in the frequency of extreme weather events such as tropical cyclones and heat waves. The intensity of tropical cyclones has been projected to increase. A general increase in extreme rainfall events are projected for southern Mozambique. Southern Africa is projected to become generally warm and increased average temperatures are projected to occur in association with increase in very hot days with maximum temperatures exceeding 35°C and associated increases in heat waves.

The key areas of agreements between different models, GCMs, statistical downscaling, and dynamical downscaling are shown in Table 2.2. All three models show an increase in projected temperatures; increase in mean minimum and maximum temperatures are indicated as consistent and robust with a minimum projected change of 0.3°C and maximum at 3.6°C. All the models show increase in very hot days or heat waves during summer months. Rainfall is consistently suggested to decrease over parts of Zimbabwe, Zambia, and western Mozambique. A decrease in winter and spring rainfall over southwestern parts of South Africa is indicated in most simulations, whereas southeastern South Africa generally receives more rainfall.

2.7 IMPACTS OF FUTURE CLIMATE SCENARIOS ON CROPS AND LIVESTOCK PRODUCTIVITY

Higher temperatures, variable rainfall, shifting seasons and extreme weather events, flooding and droughts will significantly challenge future agricultural production in most parts of southern Africa. In countries such as South Africa, which are dry, averaging 450 mm of rainfall, water is a critical issue. In (Southern Africa Development Community) SADC countries about 60–70% of water sources are used for agriculture. Global warming and associated changes in rainfall patterns are beneficial in some areas for agriculture, especially cool areas, e.g., highlands where temperatures are low for crop production (Scholes et al., 2015).

Every crop and animal has an optimal temperature optimum for growth and yield. For most crops and livestock, this is in the range of

Table 2.2 Comparison of climate change projections from the Global Circulation Models (GCMs) and the two downscaling techniques for southern Africa

	GCM	Statistical downscalings	Dynamical downscalings	Reference
Time scale	1960–2000 2030–2060	1961–2000 2036–2065	1961–2000 2036–2065	Davis (2011)
Rainfall	Decrease over central and western southern Africa	Increases over Angola, northern Mozambique and southeast South Africa during Dec., Jan., Feb. and Mar., Apr., May	Decrease in rainfall projected for western southern Africa	
	Decrease over most of southern Africa during September, October, November, and southwest South Africa during June, July, August	Decrease over Zimbabwe, Zambia, western Mozambique and parts of the southwestern coastline during Dec., Jan., Feb., and Sep., Oct., Nov.	Increases over south east South Africa	
Temperature	Increase in mean, minimum, and maximum temperature			
	1–3°C	0.8–3.6°C	0.3–3.2°C	
Extreme weather events	Increases in very hot days and heat waves	Increases in very hot days and heat waves	More extreme rainfall events over eastern southern Africa. Increase in very hot days—above 35°C	

Source: Davis, C., 2011. Climate Risk and Vulnerability: A Handbook for Southern Africa. CSIR, Pretoria. 92p. Available online at: www.rvatlas.org/SADC.

25–30°C (Scholes et al., 2015). Places where the growing season temperature is warmer than this will experience lower production as temperature rises. Whereas cooler places than optimum will benefit from warming. Mean global temperatures of up to 2°C above the preindustrial time will result in balanced food production globally. Above 3°C of warming, overall global agricultural production is projected to fall steadily. Southern Africa is warming at above the rate of global average, and mostly already above the temperature optimums. Southern Africa, therefore, falls into the category of a region where agricultural production will be negatively affected overall.

2.7.1 Crop Production

Maize and wheat production in southern Africa is projected to decrease considering direct effects above. For South Africa each 1% decline in rainfall translates to 1.1% decrease in summer maize production and a 0.05% decrease in winter wheat. Each 1°C increase in temperature results in a 5% decrease in maize and wheat production (Scholes et al., 2015). Already the climate of large swaths of Namibia, Botswana, Lesotho, and smaller proportions of Swaziland and Zimbabwe are unsuitable for maize production. The agricultural prospects of Swaziland, Botswana, Namibia, and Zimbabwe are worrying. Substantial decreases in the productivity of crops-suitable land in Namibia, Botswana, South Africa, and Zimbabwe are projected. Parts on Angola, Malawi, Mozambique, Zimbabwe, and Madagascar are also projected to experience declines in crop production under climate change. The agricultural outlook for Angola, southern Democratic Republic of Congo, Zambia, Mozambique, and Tanzania are potentially positive even under climate change. Decreasing crop production as temperature rises could be offset by improved agricultural practices, and the areas are broadly expected to experience more rainfall under climate change.

Cropping systems will tend to develop around their capacity to produce relatively higher yields and at the same time adapt to low rainfall, extreme cold, and/or hot temperatures and high evapotranspiration rates. Crop with drought, heat, and cold stress tolerance will dominate the markets and diets of people in southern Africa especially the borderline income group (those with income not significantly different from the poverty datum line). An increase in the area under irrigation will save huge amounts of money, which would otherwise go toward imports of processed and raw food products. Crops such as cassava will feature more prominently on the farms as they tolerate a range of stresses, and the produce has multiple uses. In areas where maize just survives at present there is a likelihood that no maize will be produced especially where the existing desserts will expand northwards in Botswana and Namibia and parts of Angola.

2.7.2 Livestock Production

Livestock (cattle, sheep, goats, pigs, and poultry) maintain their body temperature at between 36 and 38°C. As outside temperature approaches this fixed body temperature they must stay in shade instead of feeding, leading to less energy available for growth. Prolonged exposure to high temperatures without adequate water leads to livestock death. Heat stress is particularly a problem in dairy cows. Meat production, milk production, and fertility in livestock fall steeply when daytime temperatures exceed 30°C (Scholes et al., 2015). Therefore, the prospects of livestock production in the hot, arid parts of southern Africa under climate change look bleak.

The majority of climate models as shown in other sectors of southern Africa indicates a likely increase in average minimum and maximum temperatures. Increased temperatures impact on livestock production in a variety of ways. Heat stress can impact on food intake, fertility, live weight gain, and mortality. Different breeds have different thresholds above which they experience heat stress. Local breeds are more resistant to high temperatures. The more the temperate, breeds such as Holstein (*Bos taurus*) tend to experience heat stress above 22°C (Sanchez et al., 2009). It is clear that livestock farming under climate change in southern Africa will have to take cognizance of the existing knowledge around more traditional breeds and livestock management knowledge. Moonga and Chitambo (2010) propose that well-adapted traditional livestock breeds will play a significant role in adaptation to climate risk in southern Africa. The increased need for genetic conservation of these breeds cannot be overemphasized. However, the impacts of climate change on the livestock sector in southern Africa are an area requiring further research.

2.8 CONCLUSION

As temperature and rainfall patterns change throughout southern Africa, the areas most suitable for various crops will either expand or shrink and in some cases shift. That does not mean automatically that they will be worse off. They can adapt through developing heat or drought-tolerant varieties within a given crop or animal species. So far it has been difficult to exploit the upper temperature niches for plants or animals or exploit the water-use efficiency of plants. Where temperatures limit water use by fundamental physiological processes, even advanced technologies such as genetic modification will not help much to alter them. Agronomic practices or livestock practices, which are climate smart will be a key requirement for success. These technologies will reduce agricultural losses and at the same time limit the impacts of weeds, pests, and diseases of both crops and livestock.

REFERENCES

Alexander, L.V., Zhang, X., Peterson, T.C., Caesar, J., Gleason, B., Klein Tank, A.M.G., Haylock, M., Collins, D., Trewin, B., Rahimzadeh, F., Tagipour, A., Ambenje, P., Rupa Kumar, K., Revadekar, J., Griffiths, G., 2006. Global observed changes in daily climate extremes of temperature and precipitation. Journal of Geophysical Research 111, D05109.

Benson, C., Clay, E.J., 2000. Developing countries and the economic impacts of natural disasters. Managing Disaster Risk in Emerging Economies 11–21.

Chen, M.A., 2005. Rethinking the Informal Economy: Linkages with the Formal Economy and the Formal Regulatory Environment. DESA Working Paper No. 46. United Nations Department of Economic and Social Affairs, New York.

Chevalier, M., Chase, B.M., 2016. Determining the drivers of long-term aridity variability: a southern African case study. Journal of Quaternary Science 31 (2), 143–151.

Crowley, T.J., 2000. Causes of climate change over the past 1000 years. Science 289 (5477), 270–277.

Davis, C., 2011. Climate Risk and Vulnerability: A Handbook for Southern Africa. CSIR, Pretoria. 92p. Available online at: www.rvatlas.org/SADC.

Fauchereau, N., Trzaska, S., Rouault, M., Richard, Y., 2003. Rainfall variability and changes in southern Africa during the 20th century in the global warming Context. Natural Hazards 29, 139–154.

Fischer, G., Shah, M., Tubiello, F.N., van Velhuizen, H., 2005. Socio-economic and climate change impacts on agriculture: an integrated assessment, 1990–2080. Philosophical Transactions of the Royal Society B: Biological Sciences 360 (1463), 2067–2083.

Godfray, H.C.J., Beddington, J.R., Crute, I.R., Haddad, L., Lawrence, D., Muir, J.F., Pretty, J., Robinson, S., Thomas, S.M., Toulmin, C., 2010. Food security: the challenge of feeding 9 billion people. Science 327 (5967), 812–818.

IPCC, 2007. Climate Change 2007: The Physical Science Basis. Contribution of Working Group I to the Fourth Assessment Report of the Intergovernmental Panel on Climate Change. Cambridge University Press, New York.

Jiri, O., Mafongoya, P., Chivenge, P., 2015. The use of indigenous knowledge systems to predict seasonal quality for climate change adaptation in Zimbabwe. Climate Research 66, 103–111.

Kruger, A.C., Shongwe, S., 2004. Temperature trends in South Africa: 1960–2003. International Journal of Climatology 24, 1929–1945.

Kruger, A.C., 2006. Observed trends in daily precipitation indices in South Africa: 1910–2004. International Journal of Climatology 26, 2275–2286.

Lobell, D.B., Burke, M.B., Tebaldi, C., Mastrandrea, M.D., Falcon, W.P., Naylor, R.L., 2008. Prioritizing climate change adaptation needs for food security in 2030. Science 319 (5863), 607–610.

McMichael, A.J., Powles, J.W., Butler, C.D., Uauy, R., 2007. Food, livestock production, energy, climate change, and health. The Lancet 370 (9594), 1253–1263.

Mirza, M.M.Q., 2003. Climate change and extreme weather events: can developing countries adapt? Climate Policy 3, 233–248.

Moonga, E., Chitambo, H., 2010. The role of indigenous knowledge and Biodiversity in livestock disease management under climate change. In: ICID+18-Second International Conference: Climate, Sustainability and Development in Semi-arid Regions, August 16–20, 2010, Fortaleza, Ceará, p. 11.

Ngaira, J.K.W., 2007. Impact of climate change on agriculture in Africa by 2030. Scientific Research and Essays 2 (7), 238–243.

Pelletier, N., Tyedmers, P., 2010. Forecasting potential global environmental costs of livestock production 2000–50. Proceedings of the National Academy of Sciences 107 (43), 18371–18374.

Rosenzweig, C., Iglesias, A., Yang, X.B., Epstein, P.R., Chivian, E., 2001. Climate change and extreme weather events; implications for food production, plant diseases, and pests. Global Change & Human Health 2 (2), 90–104.

Sanchez, J.P., Miztal, I., Aguilar, I., Zumbach, B., Rekaya, R., 2009. Genetic determination of the onset of heat stress on daily milk production in US Holstein cattle. Journal of Dairy Science 92, 4035–4045.

Scholes, B., Scholes, M., Lucas, M., 2015. Climate Change: Briefings from Southern Africa. Wits University Press, Johannesburg.

Schulze, R.E., 2007. Climate Change and the Agricultural Sector in South Africa: An Assessment of Findings in the New Millennium. ACRUcons Report 55. University of KwaZulu-Natal, Pietermaritzburg.

Thornton, P.K., 2010. Livestock production: recent trends, future prospects. Philosophical Transactions of the Royal Society of London B: Biological Sciences 365 (1554), 2853–2867.

Tingem, M., Rivington, M., 2009. Adaptation for crop agriculture to climate change in Cameroon: turning on the heat. Mitigation and Adaptation Strategies for Global Change 14 (2), 153–168.

Unganai, L.S., Kogan, F.N., 1998. Drought monitoring and corn yield estimation in southern Africa from AVHRR data. Remote Sensing of Environment 63 (3), 219–232.

Vermeulen, S.J., Aggarwal, P.K., Ainslie, A., Angelone, C., Campbell, B.M., Challinor, A.J., Hansen, J.W., Ingram, J.S.I., Jarvis, A., Kristjanson, P., Lau, C., 2012. Options for support to agriculture and food security under climate change. Environmental Science and Policy 15 (1), 136–144.

Vogel, C., 1994. (Mis) management of droughts in South Africa: past, present and future. South African Journal of Science 90, 4–5.

CHAPTER 3

Advancing Key Technical Interventions Through Targeted Investment

Nhamo Nhamo[1], Kokou Kintche[2], David Chikoye[1]

[1]International Institute of Tropical Agriculture (IITA), Southern Africa Research and Administration Hub (SARAH) Campus, Lusaka, Zambia; [2]International Institute of Tropical Agriculture (IITA), Central Africa, Kinshasa, Democratic Republic of Congo

Contents

Smart Technologies for Sustainable Smallholder Agriculture
ISBN 978-0-12-810521-4
http://dx.doi.org/10.1016/B978-0-12-810521-4.00003-7

3.1 INTRODUCTION

Integrated soil fertility management (ISFM) has gained recognition throughout Africa because of its contribution to increased crop production and maintenance of soil nutrition (Vanlauwe et al., 2010; Sanginga and Woomer, 2009). This is against a background of an increasing population that has led to cultivation of marginal lands, reduction of fallowing periods, and intensified cultivation of old and inherently infertile farms in southern Africa (Mokwunye and Bationo, 2011). Land degradation has become rampant in southern Africa making agriculture unsustainable. Food and income from agriculture are threatened by both infertile soils and climate change and variability problems in southern Africa. Soil fertility is a function of climate, parent material, and management. As such, soils derived from silica-saturated igneous and sandstone-dominated lithologies form sandy soils with low inherent nutrient concentrations and limited nutrient and water-holding capacity (van Straaten, 2011). Climate variabilities and change experienced in southern Africa have challenged some of the ISFM aspects and this warrants research aimed at refinement of some of the messages on soil fertility management practices and input use.

Climate change has affected the rainfall and temperature (Oseni and Masarirambi, 2011) and this in turn has impacted on the agronomic and nutrient use efficiencies from ISFM technologies. This suggests that more efforts are needed in adapting technologies that increase productivity at farm level. Adjusting the application of the ISFM components accordingly, and deriving and targeting suitable rates of both composted manures and crop residue litter from legumes and cereals are likely to improve agronomic efficiency and nutrient cycling (Vanlauwe et al., 2015). Soil organic matter (SOM) provides source and sink for plant nutrients through the management of carbon, nitrogen, sulfur, and phosphorus cycles. However, to address the heterogeneous environments farmers work under (Tittonell et al., 2013), more systems analysis is required to support future development of climate-smart ISFM technologies. In particular the relationship between climatic parameters (e.g., temperature, rainfall patterns, and greenhouse gas emissions) and ISFM technologies in relation to developing resilient production systems in Africa needs critical analysis.

There has been a growing realization that ISM technologies cannot stand alone and deliver the food requirements of a growing population in a sustainable manner. Thus development of supporting production practices including utilization of climate forecast, high water use efficiency, and use of high-CO_2-tolerant species is required going forward. In other words, a more accommodating broad ISFM approach and model are required to ensure sustainable production systems under smart agriculture. Integrated climate-smart agricultural technologies are therefore sustainable practices and they encompass most of the ISFM practices. We define ISFM here as a suit of technologies that support key soil processes known to drive productivity and optimal management of organic resources, mineral fertility inputs, and SOM (Vanlauwe, 2004). In a holistically and sustainable manner, ISFM combines the improvement and maintenance of the production systems with emphasis on the sustainable management of soil fertility in a broader context of crop-livestock systems, i.e., it focuses on the physical, chemical, and biological attributes of soil including the social and economic values of the soil. Kimani et al. (2003) summarized ISFM principles as follows: (1) diversification of nutrient sources, (2) maximization of input use efficiency (and returns to investment), (3) knowledge-based intensive and holistic approach with long-term perspective, (4) nutrient stock and balances as indicators of productivity and sustainability, (5) interactions of biotic and abiotic factors (stresses), (6) crop livestock interactions (mixed farming), and (7) policy and institutional framework. This list largely intersects with the key attributes defining both climate-smart agriculture and sustainable agricultural approaches (Chapter 1). Although it cannot stand alone across landscape and ecosystem (natural and managed) scales, ISFM directly contributes to the broader concepts of natural resource management within the agroecosystems. The concept also has socioeconomic considerations where factors such as land tenure, input–output markets, and institutional support mechanisms to include credit facilities within the context of value chains are included (Bationo and Waswa, 2011).

The concepts of integrated inputs and efficient management for improved production at minimum environmental loss will continue to evolve and key questions to address include: how ISFM technologies will respond to crop production threats from climate extremes and the application across fragmented small farms to reclaim degraded soils and sustain production on healthy soils in southern Africa. In southern Africa, addressing the question on reclamation of degraded and fragmented land portions to support livelihoods of millions of poor farmers is absolutely critical (Carter and Murwira, 1995; Scoones, 1997). Fragmented farms are a characteristic associated with low resource endowed farmers, and the capacity of these farmers to invest in soil fertility input is minimum (Chikowo et al., 2014). For over half a decade smallholder farmers have been practicing extensive agriculture with minimum input to the soil, and relied on ash and limited organic matter in slush and burn systems (e.g., Hanjra and Culas, 2011; Holden, 1993; Matthews et al., 1992). Negative nutrient balances, soil degradation, erosion, and low response to management practice threaten future production potential on these fragmented portions of land. Technical interventions are required to reverse the negative trends.

The aim of this chapter is to discuss the contribution of ISFM to (1) the crop production potential, (2) mitigation of climate change, and (3) potential adaptation techniques to climate variability and change effects in southern Africa.

3.2 CLIMATE CHANGE, INTEGRATED SOIL FERTILITY MANAGEMENT, AND CROP PRODUCTION

Crop production in southern Africa will be constrained by both climate and soil fertility. There is evidence of climate change effects in southern Africa: crop losses, reduced number of crop growing days in a season, reduced total season or annual rainfall, number of days under flooding conditions, length of heat waves or extreme cold events including incidents of frost and increased pest and diseases. The majority of these effects impact negatively on rain-fed crop production systems on smallholder farms. Three major climate variables, i.e., rainfall, temperature, and carbon dioxide concentrations in the atmosphere, need further analysis as their effect on crop production can be resounding.

The genesis of ISFM approaches was the need to develop a holistic approach to resolving biophysical limitations and sustainably increase crop production (e.g., Nandwa and Bekunda, 1998; Palm et al., 1997a,b). Parallels can be drawn between ISFM and climate-smart agricultural approach, which despite the emphasis on climate, is a holistic approach to production, and mitigation of and adaptation to climate change. To date, ISFM offers a range of management practices that improve production and lead to adaptation to climate change, e.g., ISFM practices on diversification of nutrient sources, maximization of input use efficiency, and maintenance of balanced nutrient stock for productivity and sustainability in mixed farming (crop-livestock) systems (Vanlauwe et al., 2015). However, some of the ISFM practices are directly affected by rainfall and temperature and these need to be evaluated further given the current climate variability in southern Africa.

Crop production in southern Africa will be affected by climate and the quality of soil resources available to farmers. Across the major crops (cereals, legumes, and roots or tubers) in southern Africa the investments in germ-plasm development is beginning to yield positive results. Maize (*Zea mays* L.) has received a lot of attention because of its importance in the diets of most of the population and cultivars that fit most of the agroecological zones (short-, medium-, and long-duration cultivars are available) (Haugerud and Collinson, 1990). Similarly, legumes (soybean, groundnut, pigeon pea, cowpea, and common beans), small grains (sorghum, finger, and pearl millet), roots and tubers (cassava, potato, sweat potato), and other cereals (barley, rice, wheat) have suitable cultivars for supporting production. However, climate variability may cause shifts in rainfall patterns, create conditions suitable for pests and disease, increase CO_2 levels in the atmosphere thereby impacting C3 and C4 crops differently, and dictate a higher tolerance to extreme hot and cold temperatures (Rosenzweig and Parry, 1994; Rosenzweig et al., 2001; IPCC, 2007; Yamori et al., 2014). Crop production and development in southern Africa may need further investment to refine the current cultivars and develop new ones that will suit and adapt to evolving climate changes and variability.

3.3 ENHANCING RESOURCE UTILIZATION TO EXPLOIT SPATIAL AND TEMPORAL OPPORTUNITIES

3.3.1 Nutrient Stocks and Imbalances

Studies on nutrient stocks and balances in southern African countries suggest that macronutrients are deficient across major soil groupings, which will continue to limit yields from smallholder farms unless the problem is addressed (Nandwa and Bekunda, 1998; Stoorvogel et al., 1993). Table 3.1 shows some of stocks from work done in southern Africa. Major soil groups face increased risk of reduced nutrient stocks if no suitable site-specific, climate-sensitive, and economical fertility management practices are derived and recommended. Kihara et al. (2011), Zingore et al. (2007) reported nutrient harvest of between 19 and 196 kg N/ha, 4–17 kg P/ha, and 102–191 kg K/ha from maize trials. Without replacement of these nutrients maize production can lead to soil nutrient mining. Based on this work on evaluating nutrient stocks and status there should be enough knowledge to design the packages to reduce negative nutrient imbalances and eventually reclaim the soil. On soils with nutrient imbalances recycling attempts with no meaningful inputs added will be futile. The basis for designing balanced fertilizer regimes have been suggested by Buresh et al. (2010) (e.g., Fig. 3.1) and the characteristics of the different climatic zones can be incorporated

Table 3.1 Average N, P, and K (kg/ha year) nutrient balances of some countries in southern Africa (average for 1982–4)

Country	N	P	K
Botswana	0	1	0
Malawi	−68	−10	−44
Tanzania	−27	−4	−18
Zimbabwe	−31	−2	−22

Source: Stoorvogel, J.J., Smaling, E.A., Janssen, B.H., 1993. Calculating soil nutrient balances in Africa at different scales. Fertilizer Research 35 (3), 227–235.

Figure 3.1 An illustration of the nutrient management scheme for site-specific nutrient management. *Source: Buresh, R.J., Pampolino, M.F., Witt, C., 2010. Field-specific potassium and phosphorus balances and fertilizer requirements for irrigated rice-based cropping systems. Plant and Soil 335 (1–2), 35–64. http://dx.doi.org/10.1007/s11104-010-0441-z; Buresh, R.J., August 2010. Nutrient best management practices for rice, maize, and wheat in Asia. In: World Congress of Soil Science, vol. 16.*

by identifying suitable nutrient sources, related management practices, and step-by-step reclamation plans on dominant soil types.

Arriving at a sustainable balance would entail reducing excessive nutrient inputs and preventing deficiencies to reap agronomic, economic, and environmental benefits (Vitousek et al., 2009). Several steps will be required to reduce losses of agricultural nutrients and consequent environmental damages. In terms of agronomy, improved nutrient composition of fertilizer products, better targeted placement of nutrient inputs, and timed applications are some of the strategies that can improve nutrient use. Similarly improved crop-livestock mixtures and diets can be evaluated for implementation. Increased efforts to modify and redesign cropping systems, i.e., intercropping and rotations in multiple and monocropping arrangements and combinations of annual and perennial crops are also needed as they have the capacity to improve nutrient use efficiencies.

The major bottleneck to assess the impacts of changing farm practices at appropriate scales remains human resource capacity and infrastructure for most national agricultural institutions (Vitousek et al., 2009). This and lack of incentives to promote the adoption of nutrient-saving practices that are climate smart require both policies and investment to improve the current situation. As the ability to diagnose nutrient-driven problems improves, the design of solutions, quantification of on-farm nutrient budgets, identification of multiple pathways of inputs and losses over time, and alternative management practices will be advanced.

3.3.2 Targeting Multiple Nutrient Sources

Agricultural production is threatened by massive degradation, a narrow range of nutrient sources, and neutral to negative nutrient budgets in southern Africa (ESA) (e.g., Hartemink, 1997; Gachene et al., 1997; Nandwa, 2001). There is an urgent need for identification of suitable sources of nutrients to correct negative nutrient budgets, stabilize nutrient stocks of soils and reduce soil degradation. In resource-constrained communities, locally available organic nutrient sources are more widely used as soil amendment under smallholder farming systems (Okalebo et al., 2006; Nhamo et al., 2014). Use of organic materials guided by the information on quality indices can be enhanced by use of databases already published e.g., the organic resource database (Palm et al., 2001a,b). The practical application of such a resource contributes to the designing of appropriate ISFM practices for a range of cropping systems. Work of Nhamo et al. (2014) has shown that higher yield gains were obtained from organic fertilizer compared with mineral fertilizers in maize or rice systems. Other works, e.g., Kimani et al. (2003), Kuntashula and Mafongoya (2005), Mapfumo et al., (2005), Sileshi et al., (2007), have further highlighted the importance of quality organic matter from improved fallows, rotations, green manures, agroforestry, and farmyard manure. However, the identification of new (undocumented organic) resources and hence expansion of the organic resource database has slowed down. Given the importance of organic matter management in ISFM continued work leading to new additions to the list of new candidates organic materials fully characterized can improve resource utilization and available options for farmers practicing ISFM.

Farmyard manure, grain legumes, green manures, and *Azolla* sp. have shown positive results on yields and soil quality parameters (Kaizzi, 2002; Merteens et al., 1999). The economic consideration by smallholder farmers on biologically driven recycling of nutrients from animal manure, composts, and nitrogen fixation in legumes and mycorrhizae makes them highly acceptable (Powlson et al., 2001) as a cheaper alternative to costly mineral fertilizer inputs. Climate variability will affect the overall production of biomass, its decomposition rates will determine the emissions produced and there is need for location-specific adaptation to these changes, which emphasizes high nutrient use efficiencies and sufficient sequestration to support mitigation.

3.3.3 Manure: a Source of Emissions and Crop Nutrients

Manure is an important by-product for nutrient cycling in mixed crop-livestock farming systems in sub-Saharan Africa (SSA) (Powell et al., 2004). However, manure use can mitigate climate change on one hand while driving climate change processes on the other. Manure provides nutrients to crops thus mitigating climate change through its support for plant growth. On the other hand, manure heaps and slurries in large scale livestock systems are sources of methane, carbon dioxide and carbon monoxide, greehhouse gases which drive climate change. Livestock manure is a traditional source of nutrients for many crops (Palm et al., 2001a,b). The management of nutrient flows in crop-livestock systems (i.e., inputs and outputs; additions and loses) determines the balance ultimately (negative or positive) in the long term. The major constraint to widespread use of manure in smallholder farming systems is its limited availability, variable quality due to poor handling (Nhamo, 2011; Nzuma et al., 1998), and competing uses on the farm (Vanlauwe et al., 2003). The situation is different on large-scale commercial farms where huge manure heaps and slurries require management to avoid environmental contamination. Manure has multiple benefits when applied to the soil because it has capacity to improve the biological, physical and chemical function of the soil system, long-term quality, and addresses particular soil limitations, e.g., acidity, low CEC (cationic exchange capacity), P-fixation, and erosion. In maize systems application of an average of 5 t/ha was found to sustain modest yields (minimum 2 t/ha), supplying enough N to the crop but not P (e.g., Nhamo, 2011; Palm, 1995). Similarly, application of fertility inputs in combination with manure and lime has resulted in higher yield increases. Synergistic effects of manure–mineral fertilizer combinations were reported on smallholder farms in Zimbabwe

Table 3.2 Quality characterization and categorization used by various authors giving an indication of how manure and leguminous litter can be used as nutrient sources

Source(s)	Organic material	Cattle manure/organic material quality category		
		High	Medium	Low
Mugwira and Mukurumbira (1986)	Cattle manure	N > 1.22%	0.62–1.21% N	N < 0.62
Nhamo et al. (2003), Nhamo (2011)	Cattle manure	N > 0.9%; C:N < 20	Nd	N < 0.89%; C:N > 20; sand content > 70%
Palm et al. (1997a,b)	Organic resources for biomass transfer	N > 2.5%; lignin < 15%; polyphe-nols < 4%	N > 2.5%; lignin > 15%; polyphe-nols > 4%	N < 2.5%

Nd, not determined.
Source: Nhamo et al., 2014.

(Nhamo, 2011). Table 3.2 shows some guidelines on how to use organic materials mainly based on their chemical characteristics and quality. Raising sufficient quantities of manure on smallholder farms is in general a challenge for farmers due to livestock management numbers and practices that do not maximize manure collection. With climate variability in recent years, droughts negatively affect pastures and livestock numbers thereby reducing the quantity of available manure.

The nutrient contribution from manure and other organic materials is substantial. More work is required to determine the specific contribution of manure and other organic materials is location and cropping systems specific. Research results on composts in particular and manure in general has not been conclusive over the variability in quality and recommendations on effective use on major crops, an area which requires attention. For instance, work of Mugwira and Mukurumbira (1986), Palm et al. (2001a,b), and Nhamo (2011) characterized manure into quality categories of high, medium, and low using different indices, an issue that is yet to be resolved (Table 3.2). Recently Sileshi et al. (2017) has demonstrated the importance of element ratios (C:N and N:P) in the stoichiometry and application of animal manures. Furthermore, Nyamangara et al. (2003) studied the relationship between quantity and quality and leaching of N from manured plots. Their results from lysimeter studies showed that the risk of losing N through leaching from manure was low. The application of cattle, chicken, goat, and pig manures research results also require further analysis and development of guidelines for use. To use manure as a substitute of mineral fertilizers more data are required on nutrients supplied by manure and those taken up by crops does not lead to overapplication of some nutrients in the long term leading to higher runoff and leaching of N and P thereby polluting the environment (Edmeades, 2003).

Emissions from manure heaps require urgent redress to reduce them to lower figures. Improved processes of manure curing and storage can be improved on methane and carbon dioxide retention.

The need for decision guides on use of manure should utilize established chemical indices, the relationship between fertilizer equivalency and nutrient contents which often varies (Murwira et al., 2002) depending on handling factors such as storage and curing. In addition, guides need to reflect farmers' experience in using resources at the farm. Manure, for instance, is not a suitable source of supplying P to a high-yielding maize or rice crop because the P contents are generally low (Nhamo, 2011). Work of Chettri et al. (2003) highlighted this point, hence the need to fortifying manure for P. Probert et al. (1995) measured P ranges of 0.06% to 0.57%, a far outcry of a rice crop P requirement even where manure application rates of 37 t/ha.

3.3.4 Nutrient Recycling Using Crop Residues

Nutrient export through crop harvest without replacement is a major driver of negative nutrient balances on smallholder farms in southern Africa (Powell et al., 2004). Crop residues can be viewed as a vehicle of returning nutrients back to productive agricultural soils provided the process supports soil carbon storage. However, where residues constitute a large portion of the livestock feed and when other competing uses rely on the same, scarcity of crop residues can jeopardize their role in nutrient recycling. Burning and removal of crop residues annually are common practices among farmers mainly used to clear land for smooth passage of implements and to control weeds (Mokwunye and Bationo, 2011; Roder et al., 1998), which exacerbate mining of nutrients and directly contribute to greenhouse gas emissions and loss of soil biological properties (Wassmann and Pathak, 2007). Addition of crop residues contributes to soil organic matter continuum of fresh matter to highly decomposed forms. Direct incorporation of crop residue has been reported to cause pest and disease carried over in cropping systems; however, physical manipulation, pretreatment, or composting such residues are some of the alternatives available. The quality of most crop residues, except legumes, is low and on average a ton contributes between 6 and 23 kg N (Palm et al., 1997a,b). However, it is the changes during transformation that determine carbon sequestration in soils. The fulvic and humic acids and humin fractions have been used to characterize organic matter, the particle size and the soluble organic fractions have become more widely referred to in elation to use. No positive short-term effects of mulching on yields have been demonstrated for 3 years (Roder et al., 1998); however, mulching is important in reducing C and N losses and conservation of water.

3.3.5 Nitrogen Fixation on Smallholder Farms

Biological nitrogen fixation (BNF) contribute significantly to the systems budgets for N for both crop and livestock production systems. For crop production systems, symbiotic Bradyrhizobia and free-living *Cyanobacteria* species are the major bacterial species that are at the center of N fixation in legumes (Giller, 2001). However, Frankia and heterotrophic free-living N_2 fixers also contribute significant amounts of nitrogen to the soil system (Kahindi et al., 1997). For commonly grown legumes, Rhizobia application has been instrumental in ensuring effective N fixation and nodule effectiveness. The effectiveness of symbiosis between crop and bacteria in nodules response to both management and environmental conditions. Multifunctional legumes have a chance for adoption and can contribute to soil fertility improvement when incorporated within the cropping system. Waddington (2003), summarized work on grain and green manure legumes in southern African and as such many developments including climate change now warrant an analysis of the current issues surrounding the use of BNF technologies in southern Africa.

3.3.6 Multipurpose Legumes for Food and Soil Fertility

Legumes are clearly important crops albeit their relegation to marginal fields, smaller land allocation, limited investment in quality seed, and gender allocation on smallholder farms. They provide proteins, cash, better trade opportunities, and improve soil fertility. They are often grown as sole crops or in combination with other major crops.

The inclusion of legumes in crop sequences and combinations increases the land equivalent value, saves as a crop insurance against total loss in the case of adverse climatic conditions and nutrient sources for companion crops (Cheruiyot et al., 2001; George et al., 1994). Legumes principally contribute to the N budget through in situ fixation. Table 3.3 shows grain

Table 3.3 The nitrogen (N)-fixing potential of grain legumes and suggested use in crop production systems in Africa

Species	Common name	N-fixing potential (kg/ha)	Use in lowland rice	Source(s)
Arachis hypogaea L.	Groundnut	21–160	Rotation	MacColl (1989)
Cajanus cajan (L) Millsp.	Pigeon pea	1–163	Rotation	Becker and Johnson (1998), Mapfumo et al. (2000), Sakala et al. (2003), and Kumwenda et al. (1996)
Cicer arietinum	Chickpea	0–124	Rotation	Giller (2001)
Glycine max (L) Merr.	Soybean	26–188	Rotation	Giller (2001), Kasasa et al. (1998)
Phaseolus vulgaris L.	Common bean	0–125	Between wet and dry season	Sikombe et al. (2003)
Vigna radiata	Green gram (mung bean)	58–107	Between wet and dry season	Giller (2001)
Vigna unguiculata (L) Walp.	Cowpea	4–145	Between wet and dry season	Awonaike et al. (1990), Becker and Johnson (1998), Giller (2001), and Ahiabor and Hirata (2003)
Vigna subterranean (L) Thou.	Bambara groundnut	12–52	Rotation	Rowe and Giller (2003)

Source: Nhamo et al., 2014.

legumes commonly used in cropping systems. The potential contribution to the N budgets on the farm ranges from 105 to 206 kg/ha representing about 60–100% N_2 fixation over a period less than 120 days (Rowe and Giller, 2003). More legumes with a short biomass production cycle are needed for a range of environments. Furthermore, by growing grain legumes farmers directly benefit from the harvested grain, which can be used for food and/or to generate income when sold. Nutrient contribution and monetary returns were found to be high when cowpea or soybean was grown in crop sequences (Usman et al., 2006). A review on the economic importance of soybean in rural economies has also highlighted this point. In rural setting, legumes are a source of protein (substitute to meat products) and the various processed products have also found ready market as some of them are categorized as healthy foods for child supplements and also for adults. Inclusion of legumes in rotation system has been found to result in more N availability and hence higher grain yields on N-limited soils (Ockerby et al., 1999).

Grain legumes that have a combination of attributes such as; high grain yields, high root and shoot biomass, low N harvest index, nodulating with indigenous rhizobial strains, easy to process for food, and an available market can impact communities in climate-affected areas. However, overarching issues with legumes remain seed availability, pests and diseases, species diversity, and microsymbionts (Becker and Johnson, 1998; Nhamo et al., 2004). There is therefore high potential use of grain legumes in crop production and a range of combinations of grain legumes, cereals, and root and tuber crops under different soil and climatic conditions.

Climate change and variability will challenge both the production of grain legumes in intercrops and as sole crops. Major considerations will be on suitability of varieties to climatic zones, compatibility of crop combinations to increase water and nutrient efficiency (resources which maybe limiting with changes in climate change), and the biotic pressure exerted on legumes by climate impacts on pests and diseases.

3.3.7 Opportunities for Green Manures Use

Screening for biomass production, weed suppression, and N-fixation identified *Mucuna* sp. and *Canavalia* sp. as suitable legumes for dry season fallows (Becker and Johnson, 1998). Similar screening work is required for application of legumes in cassava and sorghum systems. Kaizzi et al. (2006) measured high biomass production from *Mucuna* ranging between 2.6 and 7.9 t/ha. Table 3.4 summarizes the N contribution of green manure

Table 3.4 Green manure legumes that can contribute nitrogen (N) through biological nitrogen fixation for soil fertility replenishment in tropical Africa

Species	N-fixing potential (kg/ha)	Author(s)	Site(s)
Canavalia ensiformis	77–256	Becker and Johnson (1998)	Cote d'Ivoire
Cassia obtusifolia L. (Sicklepod)	Nd	Kayeke et al. (2007), Kumwenda et al. (1996)	Tanzania, Malawi
Crotalaria anagyroides	23–270	Becker and Johnson (1998)	Cote d'Ivoire
Crotalaria juncea	16–43	Becker and Johnson (1998)	Cote d'Ivoire, Malawi
Crotalaria ochroleuca G. (Sunhemp)	Nd	Kayeke et al. (2007), Kumwenda et al. (1996)	Tanzania, Malawi
Lablab purpureus	7–70	Becker and Johnson (1998)	
Mucuna pruriens var. utilis	18–242	Becker and Johnson (1998), and Sanginga et al. (2001)	
Mimosa invisa L. (Colla)	Nd	Kayeke et al. (2007), Kumwenda et al. (1996)	Tanzania, Malawi
Stylosanthes guianensis	40–201	Becker and Johnson (1998)	Cote d'Ivoire
Trifolium	Nd	Amede (2003)	Ethiopia
Vicia faba	Nd		
Vigna villosa	Nd	Amede (2003)	Ethiopia

Nd, not determined.
Source: Nhamo et al., 2014.

legumes that have also been tested on lowland production systems. Up to 76 kg N/ha can be contributed by the 45- to 60-day-old green manure. The legume can utilize the residual moisture to maximize biomass production in most places (Giller, 2001). Rice relay intercropped with *Cajanus cajan* was found to be most beneficial when the legume was introduced between 28 and 56 days after rice establishment (Akanvou et al., 2001). The highly competitive *C. cajan* has been a candidate improving soil fertility through BNF due to its high biomass production (5.81–6.582 t/ha of biomass at harvest at Matopos Research Station) leading to soil fertility enhancement, improved household nutrition, and income diversity (Ncube et al., 2003). Studying the suitability of different varieties of *C. cajan* is important in determining options of fitting the legume in cropping systems of southern Africa. Although it is most appropriate to use *C. cajan* after a cereal crop in a sequence, work of Timsina et al. (1993) showed options of using cowpea on pre- and post-rice niches with a positive effect on crop yield. However, Becker et al. (1995) noted limited availability of legume seed and the labor involved as the main limitations to adoption of postseason high-yielding early-maturing legume production in rice systems. Mekuria and Siziba (2003) analysed the economic benefits and found positive payoffs from the use of *Mucuna* on maize for farmers in Malawi, Zambia, and Zimbabwe. A range of legumes are grown by farmers utilizing the temporal opportunities (postharvesting residual moisture); however, their economic value and contribution to soil fertility improvement has not been characterized to detail.

In addition to economy the legume varieties have to withstand the climate extremes in southern Africa.

Multiple purpose legumes also contribute to pest control and weed suppression (e.g., Kayeke et al., 2007). However, Balasubramanian (1999) pointed to the potential increase in bacterial leaf blight and stem borer infestation following application of green manure–rice sequences suggesting further research requirements on this matter. Equally important is to determine the major pathways through which nutrients are lost in cereal–legume systems and develop more attractive options and integrated strategies suitable for smallholder farmers (Place et al., 2003).

3.3.8 Agroforestry Systems Enhance Soil Fertility

Agroforestry is a practice which combines the growing of crops and trees on the same landscape. In recent years agroforestry systems have evolved from the Taungya systems in early reforestation programs to increased use of improved legume species in biomass transfer and in situ soil fertilization. Species selection for suitable environments has taken center stage and led to success stories. The benefits of trees in agroforestry systems include soil nutrient improvement, erosion control, wood and wood products provision, and carbon sequestration. Tree species mitigate climate change by carbon sequestration (mimicking natural forests), fertilizing the soil (increase soil organic matter and reduce nutrient loses through slow release), and providing alternative sources of wood fuel which can be regrown at a fast rate.

Work on multipurpose trees grown systematically together with crops significantly improve soil nutrient stock in particular N (Mafongoya and Nair, 1996) and provide other environmental services (Kuntashula and Mafongoya, 2005; Sileshi et al., 2007). Species attributes for successful agroforestry include quality and quantity of biomass produced on-farm for use as soil amendments. Species adaption to conditions in southern Africa and compatibility with major cropping systems are equally important. Climate change prediction suggests that droughts and flooding will be more frequent in southern Africa. As such stem nodulating agroforestry species, e.g., *Sesbania rostrata* and *Aeschynomene afraspera*, which have the capacity to continue N fixation under flooded conditions, can be utilized in lowland flooded environments for purposes of increasing N total budget on farms (Giller, 2001). Mafongoya et al. (2000) summarized the relationship between N and P mineralization and quality indicators of agroforestry tree leaves and hence their importance in the management of soil fertility. The contribution of agroforestry species to the N budget on the farm, e.g., 200 kg/ha from *Leucaena leucocephala* (Vanlauwe et al., 2012), 110 kg/ha from *S. rostrata* (Giller, 2001), 120 kg/ha from *Tephrosia vogelii* (Gathumbi et al., 2002), and an average of between 162 and 1063 kg/ha (Rowe and Giller, 2003), is substantial and can lead to large yield increases when managed efficiently. Furthermore *A. afraspera* and *A. nilotica* have been reported to accumulate 420–530 kg N/ ha under suitable growth conditions and management (Alazard and Becker, 1987). Table 3.5 summarizes a range of agroforestry species that have been studied for their N-fixing potential in recent years. However, quantification of the amount of nutrients accumulated in some species is not clear a situation which requires redress. Rather estimates have been used to get to figures of the amount of N contributed.

Biomass transfer and in situ soil fertility improvement practices have gained popularity where agroforestry trees species have been incorporated into cropping systems. Where land is limited alley cropping with agroforestry species increases returns to land. Buresh et al. (1997) described in

Table 3.5 Agroforestry species that contribute nitrogen (N) and have the potential to contribute to soil fertility replenishment in cropping systems in the tropics

Group	Legume species	N-fixing potential (kg/ha)	Use in lowland rice	Author(s)	Site(s)
2. Agroforestry species	*Acacia anguistissima*				
	Aeschynomene afraspera	59–143		Giller (2001)	Asia
	Calliandra calothyrsus				
	Dactyladenia barteri				
	Flemingia macrophylla				
	Gliricidia sepium				
	Leucaena leucocephala	76–274		Sanginga et al. (1986, 1989), Otsyina et al. (1995)	Tanzania
	Senna siamea	160	Alley cropping	Vanlauwe et al. (2012)	Nigeria
	Sesbania aculeata				
	Sesbania cannabina				
	Sesbania rostrata	83–110	Rotation and intercrop	Giller (2001)	
	Sesbania sesban	7–18	Rotation		
	Tephrosia vogelii	98–124	Rotation	Gathumbi et al. (2002)	
	Tephrosia villosa		Rotation		

Source: Nhamo et al., 2014.

detail the ideal agroforestry species for both biomass transfer and in situ improvement of soil fertility. The efficiency with which the N from prunings is utilized by the crop has been an area of interesting research for a range of environments in the 1980s and 1990s. As summarized by Ganry et al. (2001), for efficient use of agroforestry prunings it is important to consider the following factors: (1) nutrient content and release pattern of the litter, (2) seasonal rainfall patterns, (3) the soil characteristics, and (4) the Land Equivalent Ratio. Similar conclusions were also arrived at in the work of Mafongoya et al. (1997), Nair et al. (1999), and Partey et al. (2011); however, the major challenge has been to develop operational guidelines from this important scientific information applicable at farm level.

Agroforestry will play an important role in climate change interventions to reduce deforestation and land degradation.

Economic studies conducted on tree-based technologies have shown that agroforestry used together with mineral fertilizers and inoculants is profitable (Palada et al., 1992; Sanginga et al., 1988). Furthermore, N-fixing trees that can be combined with wood, edible fruits, and high protein forage for livestock have a higher chance of being used by farmers practicing mixed crop-livestock farming (Kuntashula et al., 2004).

3.3.9 Managed Weedy Fallows and Soil Rehabilitation

Fallowing was a common practice in sedentary agriculture prior to population growth and land shortages. Fallowed land often support regrowth of natural vegetation (grasses, shrubs, and trees) during a period of 1 to 5 years. Weedy fallows when given ample time can improve soil fertility and provide other ecosystem services. The success of these fallows depends on the climatic conditions, soil types, and the species composition of the regrowth. Management interventions may be an option where fallows are shorter and the species composition require improvement.

Often a weedy fallow benefits from the residual water and nutrients from the previous cash crop e.g., maize or rice. The overall effect seems to be related to the common grass species that takeover as soil cover during the fallow period. The effective contribution of these systems to nutrient and water supply to the next crop has not been reported. In a system where farmers rely on unpurchased inputs there is merit in studying the composition of weedy fallows and more importantly the proportion of leguminous weeds naturally growing on crop fields to provide options for farmers. Work of Mapfumo et al. (2005) on maize systems showed prevalence of weedy leguminous species, e.g., *Crotalaria*, *Indigofera*, and *Tephrosia* spp. Precipitation, duration of cultivation, and management regimes were major factors that explained the diversity and productivity of indigenous legumes (Mapfumo et al., 2005). In rain-fed lowlands the dry spell often mimics the aerobic upland rice ecologies and hence the relevance of weedy fallows in the production of major crops.

Work of Tauro et al. (2010) showed dominance of *Crotalaria pallida* and *Crotalaria ochroleuca* in N fixation in indigenous fallows and maize production of 3 t/ha after 2 years in managed fallows in Zimbabwe. Fallows have potential application where land is not a limiting factor, collection of legume seed and inoculating the weed fallows with seed may then become important. Although farmers have small landholdings, often a proportion is left fallow due to constraints, e.g., failure to acquire inputs and false start of a rainy season (effects of climate variability) leading to delayed planting. Furthermore, where the opportunity cost of labor is lower than the value of by-products of a system, e.g., fodders or fire wood in agroforestry systems, there exist possibilities of using fallows.

The major limitation with weedy fallows is their low capacity to produce enough biomass of high quality within a short period of time to meet the soil fertility reclamation objective of farmers. Climate variability can also limit the weed species diversity especially the legumes which can enrich the fallow through N fixation. Furthermore, there is a limit to which the weedy legumes can produce large amounts of nutrients given their environmental needs may not be met under (1) depleted soil conditions and (2) the whole range of soil types that farmers are cultivating. For instance, in soils that are flooded or under drought for a long period of time, little information is available on which legume species would thrive under such extreme conditions and the amount of biomass they can produce. There is merit in generating such climate-related information for improved management of cropping systems. Mapfumo et al. (2003) estimated a maximum biomass production of 3.2 t/ha from noncultivated herbaceous legumes when grown on a depleted acidic sandy soil with <10% clay, 0.4% organic carbon, and <3 ppm P. The continued use of fallows, which exceed the threshold population (Ganry et al., 2001) and production levels, is limited by the net nutrient mining that takes place under intensive production systems and the limited diversity of fast-growing N-fixing species; these are next generation of questions scientists need to address.

3.3.10 Azolla Contribute to Soil Fertility

The symbiosis of cyanobacteria and plants most notable in Azolla is an important group among the N_2-fixer in the tropics (Giller, 2001). The autotrophic *Anabaena azollae* species is central in the formation of Azolla (Kahindi et al., 1997). The functional roles of the different N_2-fixing species were summarized by Kahindi et al. (1997), Giller (2001). Azolla yields higher N under flooded conditions implying that its potential application and relevance is higher under irrigated rice or wheat systems. The potential application also spreads to other which can be grown in hydromorphic zones. Work of Kaizzi (2002) showed that the application of Azolla on rice field led to a significant increase in rice yield of up to 1.4 t/ha. Combining Azolla and mineral fertilizers N, P, and K increased rice yields by a further 0.9 t/ha. An economic analysis of data from the studies showed that the Azolla technology, with a cost to benefit ratio of 1.4, was as economical on rice as was the use of *Mucuna* green manure cover crop.

In a practical sense the benefit of growing azolla, a free-floating aquatic fern, is realized only on its incorporation and decomposition. Research has shown that growing Azolla both before transplanting and after transplanting and using it as a green manure have similar effects of increasing nutrients to rice crops. Research on incorporation of freshly harvested Azolla has shown that N recovery by rice crops is probably higher than in systems where Azolla is stored and processed for long periods before use. However, N losses have been reported to be lower on Azolla incorporation compared with an equivalent amount of urea fertilizers (Watanabe et al., 1989).

The work of Kushari and Watanabe (1991) showed that Azolla P requirements can limit its production hence the suggestion to split apply P on Azolla during early and later growth stages. However, there is also an observation that among the different species there are some that can tolerate P-limited conditions. Other notable constraints to Azolla growth are temperature (both high and low temperatures inhibit growth), high water requirement, and pests and diseases (Giller, 2001). More work is required and an analysis of the economics of such work relevant for climate-affected regions.

3.3.11 Fertilizer Use on Small Farms

Smallholder farmers in Africa are the least users of fertilizer inputs per unit area under agriculture. Figures suggest that application rates are still <20 kg/ha in many countries. Agricultural transformation and commercialization of crop value chains demand that application of mineral fertilizer judiciously is required to avert low yields and to prevent land degradation driven by nutrient mining.

Fertilizer use is not a silver bullet for all soil fertility-related problems. The use of mineral fertilizers has become inevitable especially on soils that are inherently poor in fertility and those that have been degraded through inappropriate farming practices (Bationo et al., 2006). Fertilizer use by smallholder producers is limited by lack of initial capital outlay and limited availability (Bationo et al., 2006; FAO, 2010; Nakano and Kajisa, 2012). In addition, inappropriate formulations and lack of site specificity lead to over- and/or underapplication of mineral nutrient elements, a technical bottleneck occasionally encountered by farmers. Nonconducive policy environment, e.g., lack of subsidies and credit facilities, discourages cash-constrained farmers from using fertilizers. Under such circumstances the rates of application of mineral fertilizer on small farms are not in line with plant and soil requirements. However, the long-term impact of disregard-ing appropriate fertilizer practices results in land degradation. In southern Africa there is high potential for fertilizer technologies to improve crop yields and reduce nutrient mining. Larson and Frisvold (1996) reported an increase in the average fertilizer rate of application from 10 to 50 kg/ha on smallholder farms. Mineral fertilizers are the most popular source of nutri-ents albeit the low application rates (Manyong et al., 2001) and high mar-ginal rate of returns, e.g., from 30 kg N/ha reported by Kajiru et al. (1998) confirming profitability of the technology. Given the need to produce more cereals and high-value crops on smaller landholdings a balanced fer-tilizer use that does not contribute to environmental degradation is required (Yunju et al., 2012).

The major weakness with the current mineral fertilizer application rates used by smallholder farmers is lack of scientific basis. On the other hand, blanket fertilizer recommendations are not supported by long-term experimental data (Okalebo et al., 2006), even where they have met crop needs in the past changes in conditions over time dictates the need for a review. Without adequate recent research results the flexibility in calculat-ing application rates to adjust for different environments and soil type is compromised. Similarly, performing robust economic analysis and systems simulation exercises is difficult without large data sets across a range of environemnts. The subsequent yields following the application of N var-ied between 1.8 and 4.4 t/ha across sites and seasons, a phenomenon com-mon on farmers' fields. Furthermore, there is paucity of information on the responses of improved crop cultivars to macronutrient elements (N, P, and K), an important factor in attaining balanced fertilization of rice for example. Pampolino et al. (2007) quantified the environmental impact and the economic benefits of applying site-specific nutrient management on cassava, maize, rice, and sorghum systems and demonstrated its merit. To arrive at increased efficiency in use of mineral fertilizers in southern Africa, a similar approach to calculating fertilizer requirements for specific climate zones is required.

Resource-constrained farmers also fail to access costly fertilizers on time for use on their farms, a factor that limits production potential. It has commonly been observed that the supply and demand dynamics at the onset of each season leads to shortages or delayed availability of fertilizer inputs. Low investment options that are attractive to smallholder farmers both on the farm and off-farm are lacking in the current cropping systems. Yunju et al. (2012) showed that with increased fertilizer use farmers tended to use more organic fertilizers, an observation that supports the important role of locally available noncash fertilizer material. We propose a strategic introduction of organic materials in a manner that will stimulate improved yield with the goal of eventually increasing fertility input, which requires cash as an intensification strategy.

Nitrogen is a major limiting element in the majority of soils and where it is deficient the symptoms are visibly evident and the resultant yield loss is huge. Vanlauwe et al. (2012) showed that N is a crucial yield determinant and that increasing its use efficiency is equally important. Similarly, there is a growing realization that maintaining adequate P, K, Ca, Mg, and Zn concentrations and preventing pH decline is important in the long-term sustainability of production systems (Vanlauwe et al., 2012). At present balanced macronutrient fertilizer application is urgently required in major cropping systems.

Since the time the estimates on nutrient balances were made for soils groups in Africa (Sanchez, 2002; Stoorvogel et al., 1993), little progress has been made in simplifying the figures into farmer-friendly decision guides aided by local characterization information. Together with data on fertilizer use strategies developed by the Food and Agriculture Organization for each country (FAO, 2010), nutrient budgets and fertilizer use and availability data can be useful pillars in developing locally appropriate mineral fertilizers techniques that are beneficial to farmers.

3.3.12 Solving Multiple Nutrient Deficiencies on Smallholder Farms

Though nitrogen is the most limiting nutrient in most soils in SSA with continuous cultivation, other macronutrient (K and P) and micronutrient (Zn, S, and B) stocks in soil are increasingly becoming limiting (Sanchez et al., 1997). Further to these the soil fabric derived from the biological and physical conditions also need attention to support the use of plant nutrients. Fertilization practices and formulations that are capable of addressing more than single nutrient deficiencies are required to maintain and increase productivity of soils. The more reason why compound fertilizers are a major feature of formulations sold for use on major crops in southern Africa. In part, ISFM has made inroad in the area of use of multiple nutrient sources in an integrated manner. Combinations of organic and mineral fertilizers result in improved uptake and an even spread of available nutrients throughout the growing season in rice (Khan et al., 2004). Combinations of 5 t/ha of cattle manure with 40 kg/ha of mineral N fertilizer on maize, for instance, statistically outyielded sole applications in Zimbabwe (Murwira et al., 2002; Nhamo, 2011). Higher synchrony of supply of N and uptake by plants largely explained the benefits from combinations of organic and mineral fertilizers. Currently, legume (cowpea and soybean), cereal (maize, rice, and sorghum), and root and tuber crop (cassava and sweet potato) production in southern Africa does not exceed the demand for food and industrial use. Use of FYM (farm yard manure), *Sesbania* sp., *Ipomea carnea*, and *Gliricidia* has been reported by Khan et al. (2004). However, more work is required on combinations covering a range of available organic resources (e.g., Table 3.6).

Low-P soils need mineral fertilizers to overcome the deficiencies, thus organic manure fortified for P can potentially be used instead. Work on phosphate rock (PR) co-composting has shown how manure curing and improving solubility of igneous phosphate rock materials can provide suitable solutions (Dhliwayo, 1999). Industrially processed Minjingu rock phosphate is another example of locally available P sources, which require innovation to tap into. The use of rock phosphate to fortify low-P organic materials or direct application has not been widely reported in the literature. In Tanzania, Kenya, and Uganda, farmers use PR (Minjingu PR products) is common but published data on such studies on the effects on most cereals are scarce. Similarly, combinations of mineral fertilizers and *Tithonia diversifolia* have shown an increase in yield (Palm et al., 1997a,b).

Palm (1995) noted that guidelines of combining organic and mineral fertility inputs have been elusive because of knowledge gaps on the links among quality of organic materials and short-term and long-term fertilizer equivalency values. Table 3.1 shows some progress in developing indices for incorporating in guidelines for use of organic materials, i.e., legumes and farmyard manures. Adequate information has been gathered into the organic

Table 3.6 Nonleguminous plants used as soil fertility amendments for improved yield in cropping systems in the tropics

Group	Legume species	N-fixing potential (kg/ha)	Use in lowland rice	Author(s)	Site(s)
3. Other	*Faidbia albida* *Tithonia diversifolia*	3–4.5 t/ha equivalent to 90 kg N/ha	Rotation or biomass transfer	Ganunga et al. (2005)	Malawi
	Senna spectabilis		Biomass transfer or in situ fertilization	Gachengo et al. (1996), Szott et al. (1991), and Garrity and Mercado (1994)	Kenya
	Chromolaena odorata *Grevillea robusta*			Palm et al. (1998) Palm et al. (1998)	

Source: Nhamo et al., 2014.

resources database (useful enough to guide a first-time user of common legumes) but there are still gaps in the development of a similar tool for use of farmyard manures. Indices used by Mugwira and Mukurumbira (1986), Nhamo et al. (2003) (e.g., Table 3.1) need to be harmonized to develop useful information on farmyard manure. Furthermore, such guidelines need to incorporate farmer perceptions and circumstances, resource allocation, soil types, and nutrient sources for them to be effective on rice.

The importance of arbuscular mycorrhizal fungi (AMF) in nutrient acquisition, cycling, water relations, bioprotection against pathogens, phytohormone production, and aggregate formation has been reported by Powlson et al. (2001), Jeffries et al. (2003). Ryan and Graham (2002) reviewed the role of AMF in crops and concluded that AMF cannot replace mineral fertilizers but there was an increase in colonization whenever soils were poor in nutrients, especially P. Where poorly soluble PR was applied, for instance, AMF colonization significantly increased, suggesting a greater role for AMF in low-input agriculture systems with high degree of internal nutrient cycling than in intensive production systems. AMF plays a role in soil health and fertility through complex microbial interactions in the rhizosphere. The success of AMF may imply changes in crop varieties, soil fertility inputs, and hence the production system and the output. There is paucity of information on how AMF technology can be effectively applied to enhance nutrient cycling in lowland and flooded systems, whereas symbiosis occurs in approximately 80% of terrestrial plants.

The application of composting techniques on manure heaps has to include two main components, one aimed at increasing the quantity and the other, quality. Co-composting manure has been reported successful with PR, for instance, to improve the P solubility and content and similarly the inclusion of litter from agroforestry and green manure cover crops can improve the quality of manure. Ganry et al. (2001) described the different classes of manure following use of different organic materials in composting. The practical relevance of this work, which has not been done across many organic materials found on the farm, is it points to opportunities that farmers can utilize directly.

3.3.13 Soil Organic Matter Management for Reduced Emissions

SOM impacts the performance of the biophysical and chemical soil system, acts as a source and sink of nutrients and carbon, and is often considered as a pillar for soil fertility management and maintenance (Nandwa, 2001; Quansah et al., 2001; Swift et al., 2004). At field level, the management of SOM varies depending on the experience, knowledge, and perceptions of farmers. Such management is often concentrated in the plow layer where shallow tillage if often applied. Work done by Giller et al. (2011) has shown that managing SOM is a big challenge especially on extremely degraded sandy soils. The results of this study challenge the soil quality management principles which revolve around application with the aim of building SOM as an important index of soil quality. The major challenges seem to be the application of SOM management on extremely degraded, nonresponsive soils and light textured soils with relatively high biological activity. Palm et al. (2001a,b) summarized the historical developments in the research on SOM principles and highlighted the progress in understanding the science driving SOM processes and later Martius et al. (2001), Vanlauwe et al. (2002) expounded on the prospects of applying the gained knowledge into practical farmer-friendly tools. From the time of these publications there has been a realization that climate extremes are an important factor to consider in designing SOM practices.

From the analysis of the evolution of SOM concepts and its relationship to sustainability of cropping systems, Manlay et al. (2007) explained the developments in research on SOM as a potential greenhouse gas sink and the challenge of intensified agricultural production on the environment. Both topics maintain relevance today as the management of SOM remains important in reducing emissions from cropping systems and in climate change mitigation strategies. It has to be noted that the value of SOM is in part exhibited when it is transformed by soil organisms; processes which lead to release of carbon dioxide and hence greenhouse gases. It would seem appropriate that with SOM a suitable conservation combined with utilization is an appropriate model as crop intensification options and soil health topics are discussed. Outrightly preservation of SOM and short-term crop intensification are apparently conflicting objectives. To harmonize the two, use of noncompetitive forms of OM management and OM-increasing site-specific nutrient management have high potential (Martius et al., 2001). This way local OM quality and quantity attributes, soil clay content, and dominant climatic parameters are taken into consideration. In addition, evaluation of trade-offs between different land-use systems and competing ones and SOM preservation will be an important milestone. We hypothesis that the management of SOM and preservation in soils reduces the sensitivity of agricultural production to climate and related risks, thereby leading to sustainable cropping systems.

The challenge going forward is to develop efficient SOM models which can minimize the emissions coming off crop intensification.

Reclamation of degraded soils has been a target of several efforts in African smallholder sector due to a long history of nutrient mining and limited soil conservation approaches.

Options for reclamation of degraded soils are often left to labor and income intensive solutions, a scenario the majority of smallholder farmers face. However, leaving such reclamation activities to a plot-by-plot efforts by individual farmers can be retrogressive as limited impact may result. A landscape approach with innovations such as combining agroforestry, value chain, soil health approaches, and management of ecosystem services emanating from these holistic practices can stand a chance. Developing alternative sources of income has been put forward as a method of reducing the pressure on the already degraded soil (Ayuk, 2001). However, farmers may not easily find these opportunities locally, leading to migration for work; if unskilled labor is provided the probability of raising enough to buy food and support the families could be difficult. Given the short-term nature of farmers' objective, investing substantial amounts of money to cover the high upfront costs associated with SOM management remains a serious challenge for farmers. We suggest a step-by-step approach to labor and cash investment on connected and jointly managed farm at catchment level as a way of increasing sustainability and stabilizing yield gains from technologies.

3.3.14 Opportunities for Phosphate Rock Utilization

Phosphate rock (PR), organic P_o, and inorganic P_i are the predominant forms available in soils for use on crops. Deposits of PR in southern Africa are dominated by the relatively insoluble apatite and secondary phosphates. Deposits are found throughout southern Africa especially in Zimbabwe (Dhliwayo, 1999), Zambia, Tanzania, and South Africa (except Lesotho and Swaziland). Fluorapatite (42% P_2O_5) is the predominant form of apatite found in indigenous P deposits. Several methods are used in processing and utilizing PR products for crop production purposes; the major methods include: phosphate beneficiation and acidulation. Other methods can be grouped into: physical (fine grinding and mechanical activation), physico-chemical (thermal processes), chemical (acidulation, partial acidulation, mixing with sulfur, heap leaching, blending and granulation, ion exchange), and biological (phospho-composting, green manuring, biosolubilization with microorganisms, use of coir dust and mycorrhizal inoculation) (Agyin-Birikorang et al., 2007; Odongo et al., 2007; Schneider et al., 2010). Direct application is also possible; however, PR dissolves faster in soils with low pH and low P and Ca. Smallholder farmers can utilize PR where the transport costs are not prohibitive.

Bationo et al. (1987) showed the benefits of direct application of PR to crops as it improved the water-soluble P in soils. Both chemical and mineralogical compositions are important attributes that determine the agronomic efficiency. In light of the predicted future scarcity of P sources PR can be explored for direct application or used as processed products by farmers (Jeng, 2011). Therefore where possible farmers can already improve the soil P capital by direct incorporation of PR products.

3.3.15 Application of Lime for pH Amelioration

Low soil pH is one of the dominant limiting factors of crop growth in most high-rainfall areas where soils are highly weathered and contain low SOM and clay content. Both nutrient availability (Lucas and Davis, 1961; Bagayoko et al., 2000) and microbial activities related to nutrient mineralization (Rousk et al., 2009) and heavy metal toxicity (e.g., Al and Mn) are directly influenced by soil pH. Aluminum toxicity is highly associated with acidic soils, which causes P fixation and deficiencies in bases. Both dolomitic and calcitic limes are commonly used as ameliorant to soil acidity. Liming is not common practice on smallholder farms as seldom soil analysis is conducted before embarking on a cropping program. Often the major limitation to application of lime is the bulky nature of the material leading to higher transport costs. However, with the advent of liquid lime formulations better crop responses to liming and affordable transport costs will be possible. On the other hand, high-pH soils, e.g., saline and sodic soils, often require physical methods of correction as well as gypsum-based chemical amendments.

3.4 THE ECONOMICS OF MAINTAINING THE DRIVERS OF ISFM'S LONG-TERM INVESTMENTS

Resolving the most critical soil fertility and related problems will require four important action points:

1. Recapitalizing soil nutrients and organic matter
2. Recycling organic matter: manures, crop residues, and domestic and industrial waste
3. Harnessing more N from BNF in mixed cropping systems
4. Soil fauna management (e.g., mycorrhiza) for deep nutrient capture.

These four are the critical areas that require investments for agricultural development to support food and incomes of farmers in southern Africa. Furthermore, SOM, mineral fertilizers, organic fertilizers, BNF and legumes, and arbuscular mycorrhiza are important pillars supporting the successful application of ISFM in a sustainable manner.

3.4.1 Are Nutrient Management Practices Profitable?

The question of profitability of input intensive agriculture in smallholder sector is of absolute importance to farmers and other value chain stakeholders.

Nutrient management has to be beneficial to farmers for sustainability. Therefore, nutrient management practices applied in isolation of the demand and supply influence on the market dynamics may not be of relevance to farmers. Combinations of mineral and organic fertilizers have the potential for contributing to profitable crop intensification agenda (e.g., Kajiru et al., 1998). With increased market organization the majority of major crops can profitably be grown using ISFM. In addition, there is an unsatisfied demand for crop-specific traits, e.g., the demand for aromatic rice in Malawi, Zambia, and Tanzania. Where ISFM options are tested it would be important to test for profitability of the combination of technologies vis-à-vis farmer circumstances. Work of Meertens et al. (1999), Meertens and Roling (2000) demonstrated the importance of a conducive environment for widespread use of mineral fertilizers. In West Africa the limited use of green manure was reported to be linked to the lack of economic incentive in managing green manures until such a time when they can be incorporated as nutrient inputs (Becker et al., 1995). It is clear that ISFM practices that are not profitable have low chances of being adopted by farmers, hence the need to revise them looking at the most suitable combinations of component technologies.

Smallholder farmers often practice extensive low-input agriculture which is linked to the traditional practices. The reasons for this practice are twofold: (1) some families never get a member to train to be a farmer and so the skills they use have been passed on to them as a tradition and (2) the management of cash flows on the farm follows a pattern of meeting the immediate needs rather than investing in the long-term sustainability of enterprises. Policy incentives may change production practices when applied together with an improved credit system on inputs (e.g., Nakano and Kajisa, 2012). Furthermore, the suggested investment interventions need to address the weak point in the crop value chains.

A substantial amount of work on value chain development in southern Africa is going on. Although valuing agrodealers, traders, stockists, and packaging and processors is undisputedly important (Webber and Labaste, 2010), there is an urgent need to reduce the cost of production through increased application of efficient technologies at farm level. Work of Kajisa and Payongayong (2011) showed irrigation infrastructure rehabilitation, chemical fertilizers, reducing labor demands, use of short-duration cultivars, farm size, and system management skills are important in increasing rice production in Mozambique. Addressing production constraints will go a long way in achieving the goals of most players in crop value chains in Africa.

3.5 RESEARCH GAPS

The current cropping systems need to evolve and incorporate ISFM principles and research is needed to address the following areas:
- Screening for suitable grain legumes, green manures, and agroforestry species.
- Fertilizer equivalency values of animal manures and crop residues.
- Fortification of organic fertilizers to improve N and P contents.
- Effectiveness of combinations of mineral and organic fertility inputs.
- Decision guide on use of combinations.
- Decomposition dynamics for organic in lowland soils.
- Working out balanced NPK and micronutrients from organic materials.
- Designing site-specific fertilizer recommendations for target yields.
- Economic viability of combinations of ISFM technologies on major cropping systems.
- ISFM is knowledge intensive and hence more innovative scaling-up methods are required for more buy-in from farmers (majority are not highly educated).

3.6 CONCLUSIONS

ISFM has advanced in recent times and numerous aspects are widely used by smallholder farmers. Major developments were around the increased integration of legumes in cropping systems. Use of both food and nonfood legumes has increased N budgets through BNF. Investment in SOM was identified as an important step in the reclamation of degraded soil and maintenance of good soil quality. Increased investment in mineral fertilizer use is expected to support crop intensification in the future. Diversified nutrient sources are important for the development of location-specific ISFM required for a range of climatic zones. Where suitable soil fertility management practices and appropriate crop cultivars are applied the impact of climate change is bound to be lower than in locations where climate variability occurs concurrently with poor soil fertility management. ISFM is important in the development of cropping systems found in southern Africa, increasing nutrient stock, reversing the effects of long-term nutrient mining, increasing soil organic matter stocks and supporting soil health. Its key attributes of replenishing soil nutrient pools, recycling of nutrients, reducing nutrient losses to the environment, and improving efficiency of external inputs need to be advanced.

Equally important is the management of the soil mineralization potential to increase long-term SOM investment in soil.

ISFM can increase yields sustainably, support adaptation, and mitigate to climate change through support of soil carbon sequestration and storage in agroforestry-based landscape management options.

REFERENCES

Agyin-Birikorang, S., Abekoe, M.K., Oladeji, O.O., 2007. Enhancing the agronomic effectiveness of natural phosphate rock with poultry manure: a way forward to sustainable crop production. Nutrient Cycling in Agroecosystems 79 (2), 113–123.

Ahiabor, B.D., Hirata, H., 2003. Associative influences of soluble phosphate, rock phosphate and arbuscula mycorrhizal fungus on plant growth and phosphorus uptake of three tropical legumes. West African Journal of Applied Ecology 4 (1), 75–90.

Akanvou, R., Kropff, M.J., Bastiaans, L., Becker, M., 2001. Evaluating the use of two contrasting legume species as relay intercrop in upland rice cropping system. Field Crops Research 74, 23–36.

Alazard, D., Becker, M., 1987. *Aeschynomene* as green manure for rice. Plant and Soil 101, 141–143.

Awonaike, K.O., Kumarasinghe, K.S., Danso, S.K.A., 1990. Nitrogen fixation and yield of cowpea (*Vigna unguiculata*) as influenced by cultivar and Bradyrhizobium strain. Field Crops Research 24 (3), 163–171.

Ayuk, E.T., 2001. Social, economic and policy dimension of soil organic matter management in sub-Saharan Africa: challenges and opportunities. Nutrient Cycling in Agroecosystems 61, 183–195.

Bagayoko, M., Alvey, S., Neumann, G., Buerkert, A., 2000. Root-induced increases in soil pH and nutrient availability to field-grown cereals and legumes on acid sandy soils of Sudano-Sahelian West Africa. Plant and Soil 225 (1–2), 117–127.

Balasubramanian, V., 1999. Farmer adoption of improved nitrogen management technologies in rice farming: technical constraints and opportunities for improvement. Nutrient Cycling in Agroecosystems 53, 93–101.

Bationo, A., Chien, S.H., Mokwunye, A.U., 1987. Chemical characteristics and agronomic values of some phosphate rocks in West Africa. In: Menyonga, J.M., Bezuneh, T., Youndeowei, A. (Eds.), Food grain production in Semi-Arid Africa. Coordination office, OAU SAFGRAD, Essex, UK.

Bationo, A., Hartemink, A., Lungu, O., Nai, M., Oko, P., Smali, E., Thiombiano, L., 2006. African soils: their productivity and profitability of fertilizer use. In: Africa Fertilizer Summit: Nourish the Soil, Feed the Continent, Abuja, p. 25.

Bationo, A., Waswa, B.S., 2011. New challenges and opportunities for integrated soil fertility management in Africa. In: Innovations as key to the green revolution in Africa. Springer, Netherlands, pp. 3–17.

Becker, M., Johnson, D.E., 1998. Legumes as dry season fallow in upland rice-based systems of West Africa. Biology and Fertility of Soils 27, 358–367.

Becker, M., Ladha, J.K., Ali, M., 1995. Green manure technology: potential, usage and limitations. A case study for lowland rice. Plant and Soil 174, 181–194.

Buresh, R.J., Sanchez, P.A., Calhoon, F., 1997. Replenishing Soil Fertility in Africa. Soil Science Society of America, Madison, WI. Special Publication No. 51.

Buresh, R.J., Pampolino, M.F., Witt, C., 2010. Field-specific potassium and phosphorus balances and fertilizer requirements for irrigated rice-based cropping systems. Plant and Soil 335 (1–2), 35–64. http://dx.doi.org/10.1007/s11104-010-0441-z.

Buresh, R.J., August 2010. Nutrient best management practices for rice, maize, and wheat in Asia. In: World Congress of Soil Science, vol. 16, .

Carter, S.E., Murwira, H.K., 1995. Spatial variability in soil fertility management and crop response in Mutoko Communal Area, Zimbabwe. Ambio 77–84.

Cheruiyot, E.K., Mumera, L.M., Nakhone, L.N., Mwonga, S.M., 2001. Rotational effects of grain legumes on maize perfomance in the rift valley highlands of Kenya. African Crop Science Journal 9 (4), 667–676.

Chettri, G.B., Ghimiray, M., Floyd, C.N., 2003. Effects of farmyard manure, fertilizers and green manuring in rice-wheat systems in Bhutan: results from long-term experiments. Experimental Agriculture 39, 129–144.

Chikowo, R., Zingore, S., Snapp, S., Johnston, A., 2014. Farm typologies, soil fertility variability and nutrient management in smallholder farming in sub-Saharan Africa. Nutrient Cycling in Agroecosystems 100 (1), 1–18.

Dhliwayo, D., 1999. Evaluation of the Agronomic Potential and Effectiveness of Zimbabwe (Dorowa) Phosphate Rock-based Phosphate Fertilizer Materials (Ph.D. thesis). University of Zimbabwe (unpublished).

Edmeades, D.C., 2003. The long-term effects of manures and fertilizers on soil productivity and quality: a review. Nutrient Cycling in Agroecosystems 66, 165–180.

FAO, 2010. Food and Agriculture Organization of the United Nations Statistical Databases–Agriculture. http://www.faostat.fao.org.

Gachene, C.K.K., Jarvis, N.J., Linner, H., Mbuvi, J.P., 1997. Soil erosion effects on soil properties in a highland area of central Kenya. Soil Science Society of America Journal 61, 559–564.

Gachengo, E., Rao, M.R., Jama, B.A., Niang, A.I., 1996. Phosphorus Release and Availability Addition of Organic Materials to Phosphorus Fixing Soils (MSc Thesis). Moi University, Eldoret, Kenya.

Ganry, F., Feller, C., Harmand, J.M., Guibert, H., 2001. Management of soil organic matter in semiarid Africa for annual cropping systems. Nutrient Cycling in Agroecosystems 61, 105–118.

Ganunga, R.P., Yerokun, O.A., Kumwenda, J.D.T., 2005. The contribution of *Tithonia diversifolia* to yield and nutrient uptake in maize in Malawi small-scale agriculture. South African Journal of Plant and Soil 22 (4), 240–245.

Garrity, D.P., Mercado Jr., A.R., 1994. Nitrogen fixation capacity in the component species of contour hedgerows: how important? Agroforestry Systems 27 (3), 241–258.

Gathumbi, S.M., Cadisch, G., Giller, K.E., 2002. 15N natural abundance assessments of N2-fixation by mixtures of trees and shrubs in improved fallows. Soil Biology and Biochemistry 34, 1059–1071.

George, T., Ladha, J.K., Garrity, D.P., Buresh, R.J., 1994. Legumes as Nitrate catch crops during dry-to-wet transition in lowland rice cropping systems. Agronomy Journal 86, 267–273.

Giller, K.E., Tittonell, P., Rufino, M.C., van Wijk, M.T., Zingore, S., Mapfumo, P., Adjei-Nsiah, S., Herrero, M., Chikowo, R., Corbeels, M., Rowe, E.C., Baijukya, F., Mwijage, A., Smith, J., Yeboah, E., van der Burg, W.J., Sanogo, O.M., Misiko, M., de Ridder, N., Karanja, S., Kaizzi, C., K'ungu, J., Mwale, M., Nwaga, D., Pacini, C., Vanlauwe, B., 2011. Communicating complexity: integrated assessment of trade-offs concerning soil fertility management within African farming systems to support innovation and development. Agricultural Systems 104, 191–203.

Giller, K.E., 2001. Nitrogen Fixation in Tropical Cropping Systems, second ed. CABI Publishing, Wallingford.

Hanjra, M.A., Culas, R.J., 2011. The political economy of maize production and poverty reduction in Zambia: analysis of the last 50 years. Journal of Asian and African Studies, 0021909611402161.

Hartemink, A.E., 1997. Soil fertility decline in some major soil groupings under permanent cropping in Tanga region, Tanzania. Geoderma 75, 215–229.

Haugerud, A., Collinson, M.P., 1990. Plants, genes and people: improving the relevance of plant breeding in Africa. Experimental Agriculture 26 (03), 341–362.

Holden, S.T., 1993. Peasant household modelling: farming systems evolution and sustainability in northern Zambia. Agricultural Economics 9 (3), 241–267.

IPCC, 2007. Climate Change: Impacts, Adaptation and Vulnerability. In: Parry, M.L., Canziani, O.F., Palutikof, J.P., van der Linden, P.J., Hanson, C.E. (Eds.), Contribution of Working Group II to the Fourth Assessment Report of the Intergovernmental Panel on Climate Change. Cambridge University Press, Cambridge.

Jeffries, P., Gianinazzi, S., Perotto, S., Turnau, K., Barea, J.M., 2003. The contribution of arbuscular mycorrhizal fungi in sustainable maintenance of plant health and soil fertility. Biology and Fertility of Soils 37, 1–16.

Jeng, A.S., 2011. African Green Revolution requires a secure source of phosphorus: a review of alternative sources and improved management options of phosphorus. In: Innovations as Key to the Green Revolution in Africa. Springer, Netherlands, pp. 123–129.

Kahindi, J.H.P., Woomer, P., George, T., de Souza Moreira, F.M., Karanja, N.K., Giller, K.E., 1997. Agricultural intensification, soil biodiversity and ecosystem functions in the tropics: the role of nitrogen-fixing bacteria. Applied Soil Ecology 6, 55–76.

Kaizzi, K.C., Ssali, H., Vleik, P.L.G., 2006. Differential use and benefits of Velvet bean (*Mucuna pruriens* var. *utilis*) and N fertilizers in maize production in contrasting agro-ecological zones of E. Uganda. Agricultural Systems 88, 44–60.

Kaizzi, K.C., 2002. The Potential Benefit of Green Manures and Inorganic Fertilizers in Cereal Production on Contrasting Soils in Eastern Uganda. Ecology and Development Series No. 4 Gottingen.

Kajiru, G.J., Kileo, R.O., Stroud, A., Budelman, A., 1998. Validating a fertilizer recommendation across a diverse cropping environment. Nutrient Cycling in Agroecosystems 51, 163–173.

Kajisa, K., Payongayong, E., 2011. Potential of and constraints to the rice Green Revolution in Mozambique: a case study of the Chokwe irrigation scheme. Food Policy 36, 615–626.

Kasasa, P., Mpepereki, S., Giller, K.E., 1998. Nodulation and yield of promiscuous soyabean (Glycine Max [L.] Merr.) varieties under field conditions. In: Waddington, S.R., Murwira, H.K., Kumwenda, J.D.T., Hikwa, D., Tagwira, F. (Eds.), Soil Fertility Research for Maize-Based Farming Systems in Malawi and Zimbabwe. Soil Fert Net and CIMMYT-Zimbabwe, Harare, pp. 99–111.

Kayeke, J., Siguba, P.K., Msaky, J.J., Mbwaga, A., 2007. Green manure and inorganic fertilizer as management strategies for witchweed and upland rice. African Crop Science Journal 15, 161–171.

Khan, A.R., Chandra, D., Nanda, P., Singh, S.S., Ghorai, A.K., Singh, S.R., 2004. Integrated nutrient management for sustainable rice production. Archives of Agronomy and Soil Science 50, 161–165.

Kimani, S.K., Nandwa, S.M., Mugendi, D.N., Obanyi, S.N., Ojiem, J., Murwira, H.K., Bationo, A., 2003. Principles of integrated soil fertility management. In: Gichuru, M.P., Bationo, A., Bekunda, M.A., Goma, H.C., Mafongoya, P.L., Mugendi, D.N., Murwira, H.K., Nandwa, S.M., Nyathi, P., Swift, M.J. (Eds.), Soil Fertility Management in Africa: A Regional Perspective. Academy of Science Publishers and Tropical Soil Biology and Fertility of CIAT, Nairobi, pp. 51–72.

Kumwenda, J.D., Waddington, S.R., Snapp, S.S., Jones, R.B., Blackie, M.J., 1996. Soil fertility management research for the maize cropping systems of smallholders in southern Africa: A review.

Kuntashula, E., Mafongoya, P.L., Sileshi, G., Lungu, S., 2004. Potential of biomass transfer technologies in sustaining vegetable production in the wetlands (dambos) of eastern Zambia. Experimental Agriculture 40 (01), 37–51.

Kuntashula, E., Mafongoya, P.L., 2005. Farmer participatory evaluation of agroforestry trees in eastern Zambia. Agricultural Systems 84 (1), 39–53.

Kushari, D., Watanabe, I., 1991. Differential responses to Azolla to phosphorus deficiency. I. Screening methods in quantity controlled conditions. Soil Science in Plant Nutrition 37, 271–282.

Larson, B.A., Frisvold, G.B., 1996. Fertilizers to support agricultural development in sub-Saharan Africa: what is needed and why. Food Policy 21 (6), 509–525.

Lucas, R.E., Davis, J., 1961. Relationships between pH values of organic soils and availabilities of 12 plant nutrients. Soil Science 92 (3), 177–182.

MacColl, D., 1989. Studies on maize (Zea mays) at Bunda, Malawi. II. Yield in short rotations with legumes. Experimental Agriculture 25 (03), 367–374.

Mafongoya, P.L., Giller, K.E., Palm, C.A., 1997. Decomposition and nitrogen release patterns of tree prunings and litter. Agroforestry systems 38 (1–3), 77–97.

Mafongoya, P.L., Nair, P., 1996. Multipurpose tree prunings as a source of nitrogen to maize under semiarid conditions in Zimbabwe. Agroforestry Systems 35 (1), 31–46.

Mafongoya, P.L., Barak, P., Reed, J.D., 2000. Carbon, nitrogen and phosphorus mineralization of tree leaves and manure. Biology and Fertility of Soils 30 (4), 298–305.

Manlay, R.J., Feller, C., Swift, M.J., 2007. Historical evolution of soil organic matter concepts and their relationship with the fertility and sustainability of cropping systems. Agriculture, Ecosystems and Environment 119, 217–233.

Manyong, V.M., Makinde, K.O., Sanginga, N., Vanlauwe, B., Diels, J., 2001. Fertilizer use and definition of farmer domains for impact-oriented research in the northern Guinea savanna of Nigeria. Nutrient Cycling in Agroecosystems 59, 129–141.

Mapfumo, P., Mpepereki, S., Mafongoya, P., 2000. Pigeonpea rhizobia prevalence and crop response to inoculation in Zimbabwean smallholder-managed soils. Experimental Agriculture, 36 (04), 423–434.

Mapfumo, P., Mtambanengwe, F., Mpepereki, S., Giller, K., 2003. Adding a new dimension to the improved fallow concept through indigenous herbaceous legumes. In: Waddington, S.R. (Ed.), Grain Legumes and Green Manures for Soil Fertility in Southern Africa: Taking Stock of Progress. Soil Fert Net and CIMMYT-Zimbabwe, Harare, pp. 67–74.

Mapfumo, P., Mtambanengwe, F., Giller, K.E., Mpepereki, S., 2005. Tapping indigenous herbaceous legumes for soil fertility management by resource-poor farmers in Zimbabwe. Agriculture, Ecosystems and Environment 109, 221–233.

Martius, C., Tiessen, H., Vlek, P.L.G., 2001. The management of organic matter in tropical soils: what are the priorities? Nutrient Cycling in Agroecosystems 61, 1–6.

Matthews, R.B., Holden, S.T., Volk, J., Lungu, S., 1992. The potential of alley cropping in improvement of cultivation systems in the high rainfall areas of Zambia I. Chitemene and Fundikila. Agroforestry Systems 17 (3), 219–240.

Meertens, H.C.C., Roling, N.G., 2000. Non adoption of rice fertilizer technology based on a farming systems research-extension methodology in Sukumaland, Tanzania: a search for reasons. Journal of Extension Systems 16, 1–22.

Meertens, H.C.C., Ndege, L.J., Lupeja, P.M., 1999. Cultivation of rainfed, lowland rice in Sukumaland, Tanzania. Agriculture, Ecosystems and Environment 76, 31–45.

Mekuria, M., Siziba, S., 2003. Financial and risk analysis to assess the potential adoption of green manure technology in Zimbabwe and Malawi. In: Waddington, S.R. (Ed.), Grain Legumes and Green Manures for Soil Fertility in Southern Africa: Taking Stock of Progress. Soil Fert Net and CIMMYT-Zimbabwe, Harare, pp. 215–221.

Mokwunye, A.U., Bationo, A., 2011. Meeting the demands for plant nutrients for an African Green Revolution: the role of indigenous agrominerals. In: Innovations as Key to the Green Revolution in Africa. Springer, Netherlands, pp. 19–29.

Mugwira, L.M., Mukurumbira, L.M., 1986. Nutrient supplying power of different groups of manure from the communal areas and commercial feedlots. Zimbabwe Agricultural Journal 83, 25–29.

Murwira, H.K., Mutuo, P., Nhamo, N., Marandu, A.E., Rabeson, R., Mwale, M., Palm, C.A., 2002. Fertilizer equivalency values of organic materials of differing quality. In: Vanlauwe, B., Diels, J., Sanginga, N., Merckx, R. (Eds.), Integrated Plant Nutrient Management in Sub-Saharan Africa: From Concept to Practice. CABI Publishing, Wallingford, pp. 113–122.

Nair, P.R., Buresh, R.J., Mugendi, D.N., Latt, C.R., 1999. Nutrient cycling in tropical agroforestry systems: myths and science. Agroforestry in sustainable agricultural systems 1–31.

Nakano, Y., Kajisa, K., 2012. The determinants of technology adoption: a case of the rice sector in Tanzania. In: Proceedings of International Association of Agricultural Economists Conference, Foz do Iguacu, Brazil, pp. 1–38.

Nandwa, S., Bekunda, M.A., 1998. Research on nutrient flows and balances in east and southern Africa: state-of-the-art. Agriculture, Ecosystems and Environment 71 (1), 5–18.

Nandwa, S.M., 2001. Soil organic carbon (SOC) management for sustainable productivity of cropping and agroforestry systems in eastern and southern Africa. Nutrient Cycling in Agroecosystems 61, 143–158.

Ncube, B., Manjala, T., Twomlow, S., 2003. Screening of short duration Pigeon pea in Matebeleland. In: Waddington, S.R. (Ed.), Grain Legumes and Green Manures for Soil Fertility in Southern Africa: Taking Stock of Progress. Soil Fert Net and CIMMYT-Zimbabwe, Harare, pp. 75–78.

Nhamo, N., Mupangwa, W., Siziba, S., Gatsi, T., Chikazunga, D., 2003. The role of cowpea (*Vigna unguiculata*) and other grain legumes in the management of soil fertility in the smallholder farming sector of Zimbabwe. In: Waddington, S.R. (Ed.), Grain Legumes and Green Manures for Soil Fertility in Southern Africa: Taking Stock of Progress. Soil Fert Net and CIMMYT-Zimbabwe, Harare, pp. 119–127.

Nhamo, N., Murwira, H.K., Giller, K.E., 2004. The relationship between nitrogen mineralization patterns and quality indices of cattle manure from different smallholder farm in Zimbabwe. In: Bationo, A. (Ed.), Managing Nutrient Cycles to Sustain Soil Fertility in Sub-Saharan Africa. Academy Science Publishers (ASP) in Association with Centro Internacional de Agricultura Tropical (CIAT), Nairobi, pp. 299–315.

Nhamo, N., 2011. Combined Organic and Inorganic Nitrogen Sources: Cattle Manure Quality, Combinations with Mineral Fertilizers and Related Yield Gains. Lambert Academic Publishing, Saarbrucken.

Nhamo, N., Rodenburg, J., Zenna, N., Makombe, G., Luzi-Kihupi, A., 2014. Narrowing the rice yield gap in East and Southern Africa: using and adapting existing technologies. Agricultural Systems 131, 45–55.

Nyamangara, J., Bergstrom, L.F., Piha, M., Giller, K.E., 2003. Fertilizer use efficiency and nitrate leaching in a tropical sandy soil. Journal of Environment Quality 32, 599–606.

Nzuma, J.K., Murwira, H.K., Mpepereki, S., 1998. Cattle manure management options for reducing nutrient losses: farmer perceptions in Mangwende. In: Waddington, S.R., Murwira, H.K., Kumwenda, J.D.T., Hikwa, D., Tagwira, F. (Eds.), Soil Fertility Research for Maize-based Farming Systems in Malawi and Zimbabwe. Soil Fert Net and CIMMYT-Zimbabwe, Harare, pp. 183–190.

Ockerby, S.E., Garside, A.L., Adkins, S.W., Holder, P.D., 1999. Prior crop and residue incorporation timing affect the response of paddy rice to fertilizer nitrogen. Australian Journal of Agricultural Research 50, 937–944.

Odongo, N.E., Hyoung-Ho, K., Choi, H.C., Van Straaten, P., McBride, B.W., Romney, D.L., 2007. Improving rock phosphate availability through feeding, mixing and processing with composting manure. Bioresource Technology 98 (15), 2911–2918.

Okalebo, J.R., Othieno, C.O., Woomer, P.L., Karanja, N.K., Semoka, J.R.M., Bekunda, M.A., Mugendi, D.N., Muasya, R.M., Bationo, A., Mukhwana, E.J., 2006. Available technologies to replenish soil fertility in East Africa. Nutrient Cycling in Agroecosystems 76, 153–170.

Oseni, T.O., Masarirambi, M.T., 2011. Effect of Climate Change on Maize (*Zea mays*) Production and Food Security in Swaziland. Change, 2, p. 3.

Otsyina, R., Hanson, J., Akyeampong, E., 1995. Leucaena in East Africa. ICRAF, Nairobi.

Palada, M.C., Kang, B.T., Claassen, S.L., 1992. Effect of alley cropping with *Leucaena leucocephala* and fertilizer application on yield of vegetable crops. Agroforestry Systems 19 (2), 139–147.

Palm, C.A., 1995. Contribution of agroforestry trees to nutrient requirements of intercropped plants. In: Agroforestry: Science, Policy and Practice. Springer, Netherlands, pp. 105–124.

Palm, C.A., Myers, R.J., Nandwa, S.M., 1997a. Combined use of organic and inorganic nutrient sources for soil fertility maintenance and replenishment. Replenishing Soil Fertility in Africa 193–217.

Palm, C.A., Myers, R.J.K., Nandwa, S.M., 1997b. Combined use of organic and inorganic nutrient sources for soil fertility maintenance and replenishment. In: Buresh, R.J., Sanchez, P.A., Calhoin, F. (Eds.), Replenishing Soil Fertility in Africa. SSSA, Madison, WI, pp. 193–218. Special Publication Number 51.

Palm, C.A., Gachengo, C.N., Delve, R., Cadisch, G., Giller, K.E., 2001a. Organic inputs for soil fertility management in tropical agroecosystems: application of an organic resource database. Agriculture, Ecosystems and Environment 83, 27–42.

Palm, C.A., Giller, K.E., Mafongoya, P.L., Swift, M.J., 2001b. Management of organic matter in the tropics: translating theory into practice. Nutrient Cycling in Agroecosystems 61, 67–75.

Palm, C.A., Murwira, H.K., Carter, S.E., 1998. Organic matter management: from science to practice. In: Waddington, S.R., Murwira, H.K., Kumwenda, J.D.T., Hikwa, D., Tagwira, F. (Eds.), Soil Fertility Research for Maize-Based Farming Systems in Malawi and Zimbabwe. Soil Fert Net and CIMMYT-Zimbabwe, Harare, pp. 21–27.

Pampolino, M.F., Manguiat, I.J., Ramanathan, S., Gines, S.C., Tan, P.S., Chi, T.T.N., Rajendran, R., Buresh, R.J., 2007. Environmental impact and economic benefits of site specific nutrient management (SSNM) in irrigated rice systems. Agricultural Systems 93, 1–24.

Partey, S.T., Quashie-Sam, S.J., Thevathasan, N.V., Gordon, A.M., 2011. Decomposition and nutrient release patterns of the leaf biomass of the wild sunflower (*Tithonia diversifolia*): a comparative study with four leguminous agroforestry species. Agroforestry Systems 81 (2), 123–134.

Place, F., Barrett, C.B., Freeman, H.A., Ramisch, J.J., Vanlauwe, B., 2003. Prospects for integrated soil fertility management using organic and inorganic inputs: evidence from smallholder African agricultural systems. Food Policy 28, 365–378.

Powell, J.M., Pearson, R.A., Hiernaux, P.H., 2004. Crop–livestock interactions in the west African Drylands. Agronomy Journal 96, 469–483.

Powlson, D.S., Hirsch, P.R., Brookes, P.C., 2001. The role of soil microorganisms in soil organic matter conservation in the tropics. Nutrient Cycling in Agroecosystems 61, 41–51.

Probert, M.E., Okalebo, J.R., Jones, R.K., 1995. The use of manure on smallholders' farms in semi-arid eastern Kenya. Experimental Agriculture 31, 371–381.

Quansah, C., Drechsel, P., Yirenkyi, B.B., Asante-Mensah, S., 2001. Farmers' perceptions and management of soil organic matter – a case study from West Africa. Nutrient Cycling in Agroecosystems 61, 205–213.

Roder, W., Keoboulapha, B., Phengchanh, S., Prot, J.C., Matias, D., 1998. Effect of residue management and fallow length on weeds and rice yield. Weed Research 38, 167–174.

Rosenzweig, C., Parry, M.L., 1994. Potential impact of climate change on world food supply. Nature 367 (6459), 133–138.

Rosenzweig, C., Iglesias, A., Yang, X.B., Epstein, P.R., Chivian, E., 2001. Climate change and extreme weather events; implications for food production, plant diseases, and pests. Global Change & Human Health 2 (2), 90–104.

Rousk, J., Brookes, P.C., Bååth, E., 2009. Contrasting soil pH effects on fungal and bacterial growth suggest functional redundancy in carbon mineralization. Applied and Environmental Microbiology 75 (6), 1589–1596.

Rowe, E., Giller, K., 2003. Legumes for soil fertility in Southern Africa: needs, potential and realities. In: Waddington, S.R. (Ed.), Grain Legumes and Green Manures for Soil Fertility in Southern Africa: Taking Stock of Progress. Soil Fert Net and CIMMYT-Zimbabwe, Harare, pp. 15–19.

Ryan, M.H., Graham, J.H., 2002. Is there a role for arbuscular mycorrhizal fungi in production agriculture? Plant and Soil 244, 263–271.

Sakala, W.D., Ligowe, I., Kayira, D., 2003. Mucuna–maize rotations and short fallows to rehabilitate smallholder farms in Malawi. In: Waddington, S.R. (Ed.), Grain Legumes and Green Manures for Soil Fertility in Southern Africa: Taking Stock of Progress, pp. 161–163.

Sanchez, P.A., 2002. Soil fertility and hunger in Africa. Science 295 (5562), 2019–2020.

Sanchez, P.A., Shepherd, K.D., Soule, M.J., Place, F.M., Buresh, R.J., Izac, A.M.N., 1997. Soil fertility replenishment in Africa. An investment in natural resource capital. In: Buresh, R.J., Sanchez, P.A., Calhoin, F. (Eds.), Replenishing Soil Fertility in Africa. SSSA, Madison, WI, pp. 1–46. Special Publication Number 51.

Sanginga, N., Ade Okogun, J., Vanlauwe, B., Diels, J., Carsky, R.J., Dashiell, K., 2001. Nitrogen contribution of promiscuous soybeans in maize-based cropping systems. Sustaining Soil Fertility in West Africa 157–177.

Sanginga, N., Woomer, P.L. (Eds.), 2009. Integrated Soil Fertility Management in Africa: Principles, Practices, and Developmental Process. Tropical Soil Biology and Fertility Institute of International Center for Tropical Agriculture (CIAT), Nairobi, p. 263.

Sanginga, N., Mulongoy, K., Ayanaba, A., 1986. Inoculation of *Leucaena leucocephala* (Lam.) de Wit with Rhizobium and its nitrogen contribution to a subsequent maize crop. Biological Agriculture and Horticulture 3, 347–352.

Sanginga, N., Mulongoy, K., Ayanaba, A., 1988. Response of Leucaena/Rhizobia symbiosis to mineral nutrient in south western Nigeria. Plant Soil 112, 121–127.

Sanginga, N., Mulongoy, K., Ayanaba, A., 1989. Nitrogen fixation of field-inoculated *Leucaena leucocephala* (Lam.) de Wit estimated by the 15N and the difference methods. Plant and Soil 117 (2), 269–274.

Schneider, K.D., Van Straaten, P., Orduña, D., Mira, R., Glasauer, S., Trevors, J., Fallow, D., Smith, P.S., 2010. Comparing phosphorus mobilization strategies using *Aspergillus niger* for the mineral dissolution of three phosphate rocks. Journal of Applied Microbiology 108 (1), 366–374.

Scoones, I., 1997. The dynamics of soil fertility change: historical perspectives on environmental transformation from Zimbabwe. Geographical Journal 161–169.

Sikombe, F., Lungu, O.I., Munyinda, K., Sakala, M., 2003. Response of bean (*Phaseolus vulgaris*, L.) cultivars to inoculation and nitrogen fertilizer in Zambia. In: Waddington, S.R. (Ed.), Grain Legumes and Green Manures for Soil Fertility in Southern Africa: Taking Stock of Progress. Soil Fert Net and CIMMYT-Zimbabwe, Harare, pp. 39–42.

Sileshi, G., Akinnifesi, F.K., Ajayi, O.C., Chakeredza, S., Kaonga, M., Matakala, P.W., 2007. Contributions of agroforestry to ecosystem services in the Miombo eco-region of eastern and southern Africa. African Journal of Environmental Science and Technology 1 (4), 68–80.

Sileshi, G.W., Nhamo, N., Mafongoya, P.L., Tanimu, J., 2017. Stoichiometry of animal manure and implications for nutrient cycling and agriculture in sub-Saharan Africa. Nutrient cycling in Agroecosystems 107 (1), 91–105.

Stoorvogel, J.J., Smaling, E.A., Janssen, B.H., 1993. Calculating soil nutrient balances in Africa at different scales. Fertilizer Research 35 (3), 227–235.

Swift, M.J., Izac, A.-M.N., van Noordwijk, M., 2004. Biodiversity and ecosystem services in agricultural landscapes – are we asking the right questions? Agriculture, Ecosystems and Environment 104, 113–134.

Szott, L.T., Palm, C.A., Sanchez, P.A., 1991. Agroforestry in acid soils of the humid tropics. Advances in Agronomy 45, 275–301.

Tauro, T.P., Nezomba, H., Mtambanengwe, F., Mapfumo, P., 2010. Population dynamics of mixed indigenous legume fallows and influence on subsequent maize following mineral P application in smallholder farming systems of Zimbabwe. Nutrient Cycling in Agroecosystems 88 (1), 91–101.

Timsina, J., Garrity, D.P., Penning, de Vries, F.W.T., Pandey, R.K., 1993. Yield stability in rice-based cropping systems: experimentation and Simulation. Agricultural Systems 42, 359–381.

Tittonell, P., Muriuki, A., Klapwijk, C.J., Shepherd, K.D., Coe, R., Vanlauwe, B., 2013. Soil heterogeneity and soil fertility gradients in smallholder farms of the East African highlands. Soil Science Society of America Journal 77 (2), 525–538.

Usman, A., Bala, A., Tiamiyu, S.A., Alabi, M.O., 2006. Use of legumes for maintenance of rice yield in lowland rice-based cropping systems in Nigeria. Journal of Sustainable Agriculture 29 (1), 43–51.

van Straaten, P., 2011. The geological basis of farming in Africa. In: Innovations as Key to the Green Revolution in Africa. Springer, Netherlands, pp. 31–47.

Vanlauwe, B., Diels, J., Aihou, K., Iwuafor, E.N.O., Lyasse, O., Sanginga, N., Merckx, R., 2002. Direct interaction between N fertilizer and organic matter: evidence from trials with 15N-labelled fertilizer. In: Vanlauwe, B., Diels, J., Sanginga, N., Merckx, R. (Eds.), Integrated Plant Nutrient Management in Sub-Saharan Africa: From Concept to Practice. CABU Publishing, Wallingford, pp. 173–184.

Vanlauwe, B., Bationo, A., Carsky, R.J., Diels, J., Sanginga, N., Schulz, S., 2003. Enhancing the contribution of legumes and biological nitrogen fixation in cropping systems: experiences from West Africa. In: Waddington, S.R. (Ed.), Grain Legumes and Green Manures for Soil Fertility in Southern Africa: Taking Stock of Progress. Soil Fert Net and CIMMYT-Zimbabwe, Harare, pp. 15–19.

Vanlauwe, B., Bationo, A., Chianu, J., Giller, K.E., Merckx, R., Mokwunye, U., Ohiokpehai, O., Pypers, P., Tabo, R., Shepherd, K.D., Smaling, E.M.A., 2010. Integrated soil fertility management operational definition and consequences for implementation and dissemination. Outlook on Agriculture 39 (1), 17–24.

Vanlauwe, B., Nziguheba, G., Nwoke, O.C., Diels, J., Sanginga, N., Merckx, R., 2012. Long-term integrated soil fertility management in South-Western Nigeria: crop performance and impact on the soil fertility status. In: Bationo, A., Waswa, B., Kihara, J., Adolwa, I., Vanlauwe, B., Saidou, K. (Eds.), Lessons Learned from Long-term Soil Fertility Management Experiments in Africa. Springer Science and Business Media, Dordrecht, pp. 175–200.

Vanlauwe, B., Descheemaeker, K., Giller, K.E., Huising, J., Merckx, R., Nziguheba, G., Wendt, J., Zingore, S., 2015. Integrated soil fertility management in sub-Saharan Africa: unravelling local adaptation. Soil 1 (1), 491.

Vanlauwe, B., 2004. Integrated Soil fertility Management Research at TSBF: the framework, principles and their application. In: Bationo, A. (Ed.), Managing Nutrient Cycles to Sustain Soil Fertility in Sub-Saharan Africa. Academy Science Publishers (ASP) in Association with Centro Internacional de Agricultura Tropical (CIAT), Nairobi, pp. 25–42.

Vitousek, P.M., Naylor, R., Crews, T., David, M.B., Drinkwater, L.E., Holland, E., Johnes, P.J., Katzenberger, J., Martinelli, L.A., Matson, P.A., Nziguheba, G., 2009. Nutrient imbalances in agricultural development. Science 324 (5934), 1519–1520.

Waddington, S.R., 2003. Grain legumes and green manures for soil fertility in southern Africa: taking stock of progress. In: Proceedings of a Conference Held 8–11 October 2002 at Leopard Rock Hotel, Vumba, Zimbabwe. Soil Fert Net and CIMMYT-Zimbabwe, Harare.

Wassmann, R., Pathak, H., 2007. Introducing greenhouse gas mitigation as a development objective in rice-based agriculture: II. Cost-benefit assessment for different technologies, regions and scales. Agricultural Systems 94, 826–840.

Watanabe, I., Ventura, W., Nascarina, G., Eskew, D.L., 1989. Fate of *Azolla* spp. and urea nitrogen applied to wetland rice (*Oryza sativa* L.). Biology and Fertility of Soils 8, 102–110.

Webber, M.C., Labaste, P., 2010. Building Competiveness in Africa's Agriculture: A Guide to Value Chain Concepts and Application. World Bank, Washington, DC.

Yamori, W., Hikosaka, K., Way, D.A., 2014. Temperature response of photosynthesis in C3, C4, and CAM plants: temperature acclimation and temperature adaptation. Photosynthesis Research 119 (1–2), 101–117.

Yunju, L., Kahrl, F., Jianjun, P., Roland-Holst, D., Yufang, S., Wilkes, A., Jianchu, X., 2012. Fertilizer use patterns in Yunnan Province, China: implications for agriculture and environmental policy. Agricultural Systems 110, 78–89.

FURTHER READING

Amede, T., 2003. Pathways for fitting legumes into East African Highlands farming systems: a dual approach. In: Waddington, S.R. (Ed.), Grain Legumes and Green Manures for Soil Fertility in Southern Africa: Taking Stock of Progress. Soil Fert Net and CIMMYT-Zimbabwe, Harare, pp. 21–29.

Balasubramanian, V., Sie, M., Hijmans, R.J., Otsuka, K., 2007. Increased rice production in sub-Saharan Africa: challenges and opportunities. Advances in Agronomy 94 (6), 55–133.

Bationo, A., Buerkert, A., 2001. Soil organic carbon management for sustainable land use in Sudano-Sahelian West Africa. Nutrient Cycling in Agroecosystems 61, 131–142.

Besmer, Y.L., Koide, R.T., Twomlow, S.J., 2003. Role of phosphorus and Arbuscular Mycorrhizal fungi on nodulation and shoot nitrogen content in groundnut and Lablab bean. In: Waddington, S.R. (Ed.), Grain Legumes and Green Manures for Soil Fertility in Southern Africa: Taking Stock of Progress. Soil Fert Net and CIMMYT-Zimbabwe, Harare, pp. 43–52.

Deb, D., Lässig, J., Kloft, M., 2012. A critical assessment of the importance of seedling age in the System of Rice Intensification (SRI) in eastern India. Experimental Agriculture 48 (3), 326–346.

Dobermann, A., 2004. A critical assessment of the system of rice intensification (SRI). Agricultural Systems 79, 261–281.

Dudal, R., 2002. Forty years of soil fertility work in Sub-Saharan Africa. In: Vanlauwe, B., Diels, J., Sanginga, N., Merckx, R. (Eds.), Integrated Plant Nutrient Management in Sub-Saharan Africa: From Concept to Practice. CABI Publishing, Wallingford, pp. 7–21.

Gilberts, R.A., 1998. Under-sowing green manures for soil fertility enhancement in the maize-based cropping systems of Malawi. In: Waddington, S.R., Murwira, H.K., Kumwenda, J.D.T., Hikwa, D., Tagwira, F. (Eds.), Soil Fertility Research for Maize-based Farming Systems in Malawi and Zimbabwe. Soil Fert Net and CIMMYT-Zimbabwe, Harare, pp. 73–80.

Giller, K.E., 2002. Targeting management of organic resources and mineral fertilizers: can we match scientists' fantasies with farmers' realities? In:Vanlauwe, B., Diels, K., Sanginga, N., Merckx, R. (Eds.), Integrated Plant Nutrient Management in Sub-Saharan Africa: From Concept to Practice. CAB International, Wallingford, pp. 155–171.

Gladwin, C.H., Thomson, A.M., Peterson, J.S., Anderson, A.S., 2001. Addressing food security in Africa via multiple livelihood strategies of women farmers. Food Policy 26, 177–207.

Haefele, S.M., Wopereis, M.C.S., Ndiaye, M.K., Kropff, M.J., 2003. A framework to improve fertilizer recommendations for irrigated rice in West Africa. Agricultural Systems 76 (1), 313–335.

Hartemink, A.E., Priess, J., Veldkamp, E., Teketay, D., Lesschen, J.P., 2005. Assessment of soil nutrient depletion and its spacial variability on smallholders' mixed farming systems in Ethiopia using partial versus full nutrient balances. Agriculture, Ecosystems and Environment 108, 1–16.

Herrera, W.T., Garity, D.P., Vejpas, C., 1997. Management of *Sesbania rostrata* green manure crops grown prior to rainfed lowland rice on sandy soils. Field Crops Research 49, 259–268.

Kihara, J., Bationo, A., Mugendi, D.N., Martius, C., Vlek, P.L., 2011. Conservation tillage, local organic resources and nitrogen fertilizer combinations affect maize productivity, soil structure and nutrient balances in semi-arid Kenya. Nutrient Cycling in Agroecosystems 90 (2), 213–225.

Kolawole, O.D., 2012. Soils, science and the politics of knowledge: how African smallholder farmers are framed and situated in the global debates on integrated soil fertility management. Land Use Policy 30, 470–484.

Krupnik, T.J., Shennan, C., Settle, W.H., Demont, M., Ndiaye, A.B., Rodenburg, J., 2012. Improving irrigated rice production in the Senegal River Valley through experiential learning and innovation. Agricultural Systems 109, 101–112.

Mariano, M.J., Villano, R., Fleming, E., 2012. Factors influencing farmers' adoption of modern rice technologies and good management practices in the Philippines. Agricultural Systems 110, 41–53.

Menete, M.Y.L., van Es, H.M., Brito, R.M.L., DeGloria, S.D., Famba, S., 2008. Evaluation of system of rice intensification (SRI) component practices and their synergies on salt-affected soils. Field Crop Research 109, 34–44.

Meertens, H.C.C., Kajiru, G.J., Ndege, L.J., Enserink, H.J., Brouwer, J., 2003. Evaluation of on-farm soil fertility research in the rainfed lowland rice fields of Sukumaland, Tanzania. Experimental Agriculture 39, 65–79.

Mitra, S., Wassmann, R., Jain, M.C., Pathak, H., 2002. Properties of rice soils affecting methane production potentials: 1. Temporal patterns and diagnostic procedures. Nutrient Cycling in Agroecosystems 64, 169–182.

Myers, R.J.K., Palm, C.A., Cuevas, E., Gunatilleke, I.U.N., Brossard, M., 1994. The synchronization of nutrient mineralization and plant nutrient demand. In: Woomer, P.L., Swift, M.J. (Eds.), The Biological Management of Tropical Soil Fertility. John Wiley and Sons, Chichester, pp. 81–116.

Otsuka, K., Kalirajan, K.P., 2005. An overview and prospects for a green revolution in sub-Saharan Africa. Journal of Agriculture and Development Economics 2, 1–6.

Peter, G., Runge-Metzger, A., 1994. Monocropping, intercropping or crop rotation? An economic case study from the west African Guinea Savannah with special reference to risk. Agricultural Systems 45, 123–143.

Rodenburg, J., Johnson, D.E., 2009. Weed management in rice-based cropping systems in Africa. Advances in Agronomy 103, 149–218.

Roose, E., Barthes, B., 2001. Organic matter management for soil conservation and productivity restoration in Africa: a contribution from Francophone research. Nutrient Cycling in Agroecosystems 61, 159–170.

Seck, P.A., Tollens, E., Wopereis, M.C.S., Diagne, A., Bamba, I., 2010. Rising trends and variability of rice prices: threats and opportunities for sub-Saharan Africa. Food Policy 355, 403–411.

Sheldrick, W.F., Lingard, J., 2004. The use of audits to determine nutrient balances in Africa. Food Policy 29, 61–98.

Sistani, K.R., Reddy, K.C., Kanyika, W., Savant, N.K., 1998. Integration of rice crop residue into sustainable rice production system. Journal of Plant Nutrition 21 (9), 1855–1866.

Stoop, W.A., Uphoff, N., Kassam, A., 2002. A review of agricultural research issues raised by the System of Rice Intensification (SRI) from Madagascar: opportunities of improving farming systems for resource poor farmers. Agricultural Systems 71, 249–274.

Stoop, W.A., Adam, A., Kassam, A., 2009. Comparing rice production systems: a challenge for agronomic research and for the dissemination of knowledge-intensive farming practices. Agricultural Water Management 96, 1491–1501.

Styger, E., Ag Atter, M., Guindo, H., Ibriham, H., Diaty, M., Abba, I., Traore, M., 2010. Application of system of rice intensification practices in the arid environment of the Timbuktu region in Mali. Paddy and Water Environment 9, 137–144.

Thornton, P.K., van de Steeg, J., Notenbaert, A., Herrero, M., 2009. The impact of climate change on livestock and livestock systems in developing countries. A review of what we can know and what we need to know. Agricultural Systems 101, 113–127.

Timsina, J., Connor, D.J., 2001. Productivity and management of rice and wheat cropping systems: issues and challenges. Field Crops Research 69, 93–132.

Tittonell, P., Giller, K.E., 2012. When yield gaps are poverty traps: the paradigm of ecological intensification in African smallholder agriculture. Field Crops Research. http://dx.doi.org/10.1016/j.fcr.2012.10.007.

Tsujimoto, Y., Horie, T., Randriamihary, H., Shiraiwa, T., Homma, K., 2009. Soil management: the key factor for higher productivity in the fields utilizing the system of rice intensification SRI in the central highlands of Madagascar. Agricultural Systems 100, 61–71.

Vanlauwe, B., Bationo, A., 2003. Enhancing the contribution of legumes and biological nitrogen fixation in cropping systems: experience from West Africa. In: Waddington, S.R. (Ed.), Grain Legumes and Green Manures for Soil Fertility in Southern Africa: Taking Stock of Progress. Soil Fert Net and CIMMYT-Zimbabwe, Harare, pp. 3–13.

Witt, C., Buresh, R.J., Balasubramanian, V., Dawe, D., Dobermann, A., 2004. Principles and promotion of site-specific nutrient management. In: Dobermann, A., Witt, C., Dawe, D. (Eds.), Increasing Productivity of Intensive Rice Systems through Site-Specific Nutrient Management. Science Publishers, Inc. IRRI, Enfield, NH, USA and Los Banos, Philippines, pp. 397–410.

Woomer, P.L., Swift, M.J., 1994. The Biological Management of Tropical Soil Fertility. John Wiley and Sons, UK.

Wopereis, M.C.S., Donovan, C., Nebie, B., Guindo, D., N'-Diaye, M.K., 1999. Soil fertility management in irrigated rice systems in the Sahel and Savanna regions of West Africa. Part I. Agronomic analysis. Field Crop Research 61, 125–145.

Zandsta, H.G., 1979. Cropping system research for the Asian rice farmer. Agricultural Systems 4, 135–153.

Zingore, S., Murwira, H.K., Delve, R.J., Giller, K.E., 2007. Influence of nutrient management strategies on variability of soil fertility, crop yields and nutrient balances on smallholder farms in Zimbabwe. Agriculture, Ecosystems and Environment 119 (1), 112–126.

CHAPTER 4

Exploring Climatic Resilience Through Genetic Improvement for Food and Income Crops

Martin Chiona[1], Godfrey Chigeza[2], Pheneas Ntawuruhunga[2]
[1]Zambia Agricultural Research Institute (ZARI), Mansa, Zambia; [2]International Institute of Tropical Agriculture (IITA), Southern Africa Research and Administration Hub (SARAH) Campus, Lusaka, Zambia

Contents

4.1 INTRODUCTION

The evidence of climate change is overwhelming with the vulnerable people from poorest countries suffering the most. Climatic change is having and will continue to have drastic impact on agronomic conditions including temperature, precipitation, soil nutrients, and the incidence of disease pests, to name a few. Food security has become thus critical for smallholder farmers in the southern Africa region (Lobell et al., 2008; Crush and Frayne, 2010) where about 70% of the population depend on agriculture for food, income, and employment. To face this looming threat, significant progress in developing new breeding strategies has been made over the last few decades in developed countries but not at the same pace in developing countries such as in the southern Africa region where they are needed the most.

One of the effective ways for crop production to grow or at least to stay stable under new challenges from climate change is through the use of resilient improved varieties. The genetic diversity of crop plants is the foundation for the sustainable development of new varieties for present and future challenges. Resource-poor farmers have been using genetic diversity intelligently over centuries to develop varieties adapted to their own environmental stress conditions some of which are still in use.

Plant breeding has been developing varieties for heat, drought, and flood stresses, but with more severe and frequent challenges from aggravated climate change needs extra immediate support to overcome the challenges. Plant breeding can develop varieties to cope with climate change through many different techniques ranging from simply selecting plants in farmers' fields with desirable traits for end users to more complex classical and molecular techniques that speed the process. With the discovery of genetics, plant breeding has become in-plant breeding, and the development of new adapted varieties has become a more precise and rapid process. The science supporting plant breeding is advancing rapidly, and with sustained support, plant breeding will thus make an even greater contribution to feeding the world and tackling climate change.

Among agricultural interventions, improved germplasm stands out as one of the pillars through which agricultural transformation in Africa will continue to depend on for the unforeseeable future. Although genetic improvement for yield has been achieved immensely in a number of agricultural crop commodities with varieties adapted to different to agroecological regions at varying levels, the impact of climate change threatens to reverse these gains especially in Sub Saharan Africa (SSA). Compared to other regions, the impact of climate change on agricultural production in SSA tends to be higher due to lower precipitation and higher baseline temperatures (Kotir, 2011). This is exuberated by the worn-out soils that are inherently poor in organic matter due to limited microbial activities among others. Uncertainties in yields as a result of climate change remain high for a number of crops, making crop breeding a never-ending challenge of developing new multiple stress-tolerant crop varieties for diverse spectrum of biotic and abiotic stress conditions, e.g., diseases, pests, low soil fertility, and drought. The challenge of developing multiple stress-tolerant genotypes based on climatic change is the inability of scientists to create future ecosystems to evaluate or phenotype the desired traits.

The increased efforts in breeding for higher yield, low soil fertility, drought, extreme temperatures, and disease incidents are beginning to pay dividend in a significant way. In the past 20 years, the science of germplasm development has introduced advanced methods of breeding leading to reduced incubation periods for crop cultivars. For example, the molecular marker–assisted breeding techniques including genomewide are becoming very important in identifying and managing the genetic make-up of crop varieties. Further molecular markers are used to identify similarities and common functional genes along the chromosomes. It is envisaged that the new breeding approaches will contribute immensely to the development of climate smart germplasm, which will in turn improve resilience of crops during production. In this chapter, climate resilience of crops such as soybean, maize and cassava are taken as example for which the breeding efforts are discussed and future models of applying crop improvement techniques explored.

4.2 PROGRESS IN DEVELOPING GENETIC MATERIALS SUITABLE FOR THE ENVIRONMENTAL CONDITIONS IN SOUTHERN AFRICA

A number of crop varieties currently under cultivation in southern Africa have been developed through the conventional breeding methods. The methods involved selecting parents and isolating them to allow natural pollination by bees or other insects resulting in variability being generated that later becomes the subject of selection. This is particularly true for cross-pollinated crops such as cassava. The result is half-sib-progenies that are exposed to various environments of interest to determine the stability of the traits being considered. Full-sib-progenies are also generated through hand pollination. The efforts made to improve cassava yield are generally not geared toward the highest possible yield under favorable conditions, but rather toward obtaining stable yields in marginal conditions where cassava is grown at present (Legg et al., 2015) and this experience will likely be useful under these evolving climate conditions. For a crop such as maize, female parents are detassled to avoid self-pollination that may render the progeny not benefitting from the genes of interest from other parents. These scenarios where both parents are known the full sibs are produced and selections made from them. In self-pollinated crops, e.g., soybean, hand pollinations are done to move pollen from one parent to the stigma of another hence facilitating pollination and subsequently fertilization to produce new progeny. Hand pollination is practiced in cross-pollinated crops too to generate variability for selection. Almost all the current crop varieties, except for a few resulting from modern biotechnology, have been developed through these breeding processes. Conventional breeding has resulted in the identification of crop varieties that are resilient in coping with adverse climatic conditions. However, the process is tedious as there is no way of establishing whether the traits of interest are available in the progeny before they are exposed to the test environment. Hence, it is possible for one to make crosses in their lifetime without achieving the desirable results.

4.2.1 Soybean

The empirical evidence of increased crop yields through the use of new crop varieties in conjunction with good agronomic practices as management tools in crop production clearly demonstrate the impact of breeding-related technologies in supporting the ever increasing demand for food, feed, and fiber. Fehr (1984) estimated the contribution of plant breeding to crop productivity to be around 50%, and Rowntree et al. (2013) also found that about half of the yield gain in soybean can be attributed to genetic improvement. Studies have shown that the minimum contribution of genetic improvement in yield for cereals and oil crops since 1982 was estimated to be 88% with negligible contribution from agronomic changes (Mackay et al., 2011). In soybean, Tefera et al. (2010) showed that a yield increased from 1117 to 1710 kg/ha during two decades of breeding resulting in an average rate of increase of 24.2 kg/ha per year or 2.2% per year per release period between 1980 and 1996. Recent production (Food and Agriculture Organization) FAO data (2003–14) indicates that yield of soybean is increasing in some countries, however, in others yields have either stabilized or decreased, e.g., in countries such as Nigeria and Zimbabwe (Table 4.1 and Fig. 4.1). The decrease in yield might be due to a number of factors, which include climate change, rate of varietal replacement, and national agricultural policies. In Zambia, the annual increase in yield between 2003 and 2014 has been 82.5 kg/year, mainly because of the new varieties from the private sector and the production practices, which currently include supplementary irrigation. The yields under smallholder farming communities are low in Zambia with an average yield in the range of 1.2 tons/ha.

Excluding South Africa, to date there are 194 soybean varieties released in 16 African countries of which 82 were developed by (International Institute of Tropical Agriculture) IITA (Tables 4.2). Most of the varieties developed by IITA are promiscuous nodulating, which require less inputs and yields more

Table 4.1 Yield increases of soybean in selected countries

Country	5-Year rolling mean (tons/ha)	Yield increase (kg/ha/yr)	R-squared
USA	2.92	33.80	0.34
Brazil	2.90	45.90	0.36
Zambia	2.02	82.50	0.74
Zimbabwe	1.36	−24.40	0.47
Nigeria	0.96	−0.70	0.00
Malawi	1.00	37.10	0.75

FAO statistics 2004–14.

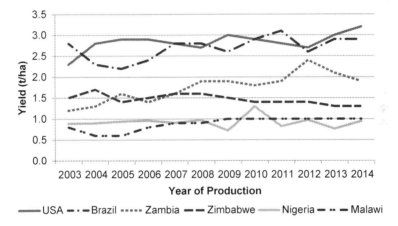

Figure 4.1 Yield increase (tons/ha) of soybean under farmers condition (FAO data, 2016).

than 1.5 tons/ha grain and more than 2 tons/ha fodder yields under farmers' conditions, which is more than threefold the yields farmers were getting earlier. In addition, the IITA varieties were bred for improved seed longevity, resistance to pod shattering, lodging, and important diseases such as rust, bacterial pustule, and frogeye leaf spot. The varieties have the ability to tolerate low levels of available phosphorus in the soil. Phosphorus is an essential element for nitrogen fixation in soybeans. The nitrogen fixing capacity of soybean is a great advantage to farmers because nitrogen is often the most limiting nutrient in the highly weathered tropical soils.

4.2.2 Cassava Climate Change and Variability

IITA's cassava breeding program in collaboration with various national program partners deployed important germplasm in the region in response

Table 4.2 Number of soybean varieties released in SSA during the past five decades of breeding

Country	IITA varieties	Non-IITA varieties	Total
Benin	12	3	15
Burundi	3	7	10
Cameron	9	11	20
Cote d'Ivoire	15	3	18
DR Congo	6	0	6
Ethiopia	3	0	3
Ghana	12	1	13
Kenya	4	6	10
Malawi	1	15	16
Mozambique	5	0	5
Nigeria	16	4	20
Tanzania	2	6	8
Togo	6	1	7
Uganda	4	8	12
Zambia	2	26	28
Zimbabwe	0	31	31
Total*	100	122	222
Total without duplicates	82	112	194

*Total does not include South Africa.

to the (Southern Africa Development Community) SADC region's call to mitigate the effect of drought. Examples of regional impact of this collaboration have been shown in Zambia and Malawi. For instance, the Malawi's Root and Tuber Crops Research Program, established in 1978, in collaboration with IITA's cassava research breeding programs focused on identification of best local variety and distribution of clean planting material to avoid pest contamination. In the 1980s, the program released a first wave of improved local varieties, including *Gomani* and *Mbundumali*, which were selected for early bulking properties and tolerance to mosaic virus disease, a disease transmitted by whiteflies and infected cuttings (Arega et al., 2013). The yield gains estimated in 2013 comparing the use of improved varieties and the clean planting materials in Malawi were 55% and 33%, respectively. Concerted seed multiplication efforts saw massive cassava production throughout the country (Benesi et al., 1995). Using the conventional plant breeding methods more improved high yielding, pest, and disease-resistant varieties adapted to different agroecologies were developed (Tables 4.3 and 4.4) and have been released, multiplied, and distributed to farmers with efforts complemented by different donor development support projects (Ntawuruhunga et al., 2011).

In as much as crop varieties that are climate smart have been developed, their availability on farmers' fields is limited due to absence, faulty, or lack of seed systems. This is particularly true for self-pollinated and vegetatively propagated crops. The situation is different for cross-pollinated varieties such as maize where hybrids are produced by seed companies who take the challenge of multiplying the seed at a profit. Seed systems for quick response to threatening stresses for root crops such as cassava require a very well thought effective approach to deliver the improved varieties to farmers. New varieties with increased diversity have to be generated by breeders at a pace that meets changing demand and requirement due to climate change. Of course, this will necessitate efficient methodology in phenotyping and genotyping platforms. Appropriate institutional framework, policies, and practices that support the delivery seeds systems are a must.

Seed systems are normally organized in different ways depending on commodity in each country. Seed systems can be complicated or simple and local. Generally smallholder farmers rely mostly on informal seed systems and have little access to commercial seeds mostly because of lack of money or nonavailability of seeds, particularly those of root crops such as cassava. Due to farmer heterogeneity, both commercially produced seeds from registered companies and informal farmer-led seed systems, e.g., for root crops will be essential in the distribution of climate change-adapted materials.

Institutional framework that will establish linkage between local communities and research institutions will be necessary to maintain quality and sanity. At the regional level, unharmonized seed regulations remain a barrier to seeds exchange and trade between countries. Meanwhile we foresee that climate change will bring more extreme challenges that will intensify pressure on seed security. Facilitating seed exchange among countries will be necessary to cope with seed shortage in some countries due to climate change. The (Common Market for Eastern and Southern Africa) COMESA policy of harmonization of seed regulatory framework at the subregional and regional levels will be a key to facilitate administrative procedures for cross-border seeds trade and exchange.

To avoid exacerbation of virus diseases spread there should be facilities and capacity to produce clean indexed planting materials for root crops, cassava in particular. In the same vein the establishment of regional variety release and variety catalog will help to bring a diversity of varieties with huge potential to adapt to climate change in the region.

Table 4.3 Yield and quality characteristics of improved cassava varieties released in Malawi between 1980 and 2009

Variety	Category	Release year	IITA material used	Yield (tons/ha)	Maturity (MAP)	Major attributes	
						Taste	Disease tolerance
Manyokola	Local selection	1980s	None	25	9–15	Sweet	Tolerates CGM, CBSD but is susceptible to CMD
Gomani	Local selection	1980s	None	25	9–12	Bitter	Susceptible to CGM, CBSD, CMD
Mkondezi (MK91/478)	Improved	1999	Seed population	40	9–15	Bitter	Tolerates CMD and CM
Maunjili (TMS 91934)	Improved	1999	IITA introduction	35	9–12	Bitter	Tolerates CMD, CM, and CGM
Silira (TMS 60142)	Improved	1999	IITA introduction	25	12–15	Bitter	Tolerates CMD and CM but is susceptible to CGM
Sauti (CH92/077)	Improved	2002	IITA seed population	25	12–15	Bitter	Tolerates CMD, CM, and CGM
Yizaso (CH92/112)	Improved	2002	IITA seed population	25	12–15	Bitter	Tolerates CMD, CM, and CGM
Phoso (LCN 8010)	Improved	2008	IITA introduction	35	9–15	Bitter	Tolerates CMD and CBSD
Mulola (TMS 83350)	Improved	2008	IITA introduction	40	9–15	Bitter	Tolerates CMD, CM, and CGM
Sagonja (CH92/082)	Improved	2009	IITA seed population	25–35	9–15	Bitter	Tolerates CMD, CBSD, CM, and CGM
Chiombola (TME 6)	Improved	2009	IITA introduction	45	9–15	Bitter	Tolerates CMD and CGM

CBSD, Cassava brown streak disease; CM, cassava mealy bug; CGM, cassava green mite; CMD, cassava mosaic disease; LCN, low cyanogen cassava; TME, Tropical Manihoti Esculenta; TMS, tropical manioc selection.

Source: Arega, A.D., Khataza, R., Chibwana, C., Ntawuruhunga, P., Moyo, C., 2013. Economic impacts of cassava research and extension in Malawi and Zambia. Journal of Development and Agricultural Economics 5 (11), 457–469.

Table 4.4 Yield and quality characteristics of improved cassava varieties released in Zambia, 1990–2000

Variety	Category	IITA material	Year of release	Yield (tons/ha)	Maturity (MAP)	Taste	DMC (%)	Reaction to pest/diseases
Bangweulu	Local selection	None	1993	31	12–16	Bitter	39	Moderately resistant
Kapumba	Local selection	None	1993	22	16–24	Sweet	45	Moderately resistant
Nalumino	Local selection	None	1993	29	16–24	Bitter	46	Resistant
Mweru	Improved	IITA male × Nalumino	2000	41	16	Sweet	42	Tolerant
Chila	Improved	IITA male × Nalumino	2000	35	16	Bitter	41	Moderately tolerant
Tanganyika	Improved	IITA male × Nalumino	2000	36	16	Sweet	41	Moderately tolerant
Kampolombo	Improved	IITA male × Nalumino	2000	39	16	Sweet		Moderately tolerant

DMC, dry matter content.
Source: Haggblade, S., Nyembe, M., 2008. Commercial dynamics in Zambia's cassava value chains. Food Security Research Project, Lusaka, Zambia.

4.2.3 Maize

Extreme temperatures during the growing season are a challenge for maize production and a major consequence of climate changes. Therefore, breeding efforts focused on improving tolerance to abiotic stresses and is desirable and currently needed to develop cultivars stable to environmental changes.

For example, the yield and quality losses are maximized when drought occurs during flowering time in maize. In the absence of stress, tolerance mechanisms often have a significant grain yield penalty but a stability advantage. The use of reliable locations where stress intensity can be managed has made significant improvement for breeding purposes. Reliable drought locations increase the accuracy of genetic estimates in which drought-associated traits can be measured (Bolanos and Edmeades, 1996; Carena, 2005). Drought tolerance is genetically a very complex trait with large genotype by environment interactions. Therefore, the marker-assisted selection and transgenic approaches are challenging. Nontransgenic breeding approaches have a potential of significantly increasing genetic progress exploiting polygenic effects compared to transgenic approaches exploiting single-gene effects.

4.3 MODERN BREEDING TECHNIQUES FOR MAJOR CROPS IN AFRICA: MAIZE, SOYBEAN, AND CASSAVA

Although conventional breeding has been known to produce desirable results, there is, however, no way of determining whether a particular progeny has a gene for the trait of interest without exposing it to the test environment. Hence, it is possible to continue working with materials supposedly having the desirable genes only to discover years after the crosses were made that the genes that express a particular trait is absent. By then, time and resources will have been wasted and the process will have to be started again. The efficiency of these traditional methods has been questionable. Therefore, modern advanced breeding techniques can help the breeder ascertain, early in the breeding program, whether a particular progeny of interest has the required gene that expresses the trait of interest. This is done by extracting DNA and running a PCR to increase the DNA fragments of interest and testing it against the standard; in the laboratory. Progenies not having the gene for the trait of interest can be discarded immediately to avoid the cost of maintaining them. This process is referred to as marker assisted selection. Progenies that have the genes of interest are then exposed to different environments to check on their expression in the environment of interest. Some genes do not express in some environments due to $G \times E$ interaction. The ones that express the trait are selected for further evaluation.

Ceballos et al. (2015) consolidated relevant molecular and phenotypic information on cassava to demonstrate the relevance of heterosis, and alternatives to exploit it by integrating different tools. The most challenging problem in cassava breeding is how to identify the best genotype out of the billions that two (heterozygous) progenitors could hypothetically produce. They discussed how new technologies such as genomic selection, use of inbred progenitors based on doubled haploids, and induction of flowering can be employed for accelerating genetic gains in cassava. They demonstrated the potential impact of using different technologies for maximizing gains for key traits in cassava and highlighted the advantages of integrating them.

Where a gene for a particular trait is missing in a progeny of interest, the missing gene fragment can still be inserted on the specific chromosome of the progeny. The gene expression can still be verified by exposing the progeny to the test environment. With this kind of capability, considerable progress in identifying crop varieties that are resilient to and/or can tolerate adverse weather conditions can be achieved in a short period. Modern advanced breeding techniques are very expensive and one needs to weigh the benefits against the investment in the technology.

4.4 BREEDING FOR TARGET ENVIRONMENTS AND EXTREMES OF WEATHER AND CLIMATE

Breeding for target environments entails testing progenies from crosses in the target environments where the varieties are intended to be grown. The exposure of the progeny could either be in the actual environment or simulated environment in the green house. The progenies that tolerate the extremes of whether and climate are selected for further evaluation.

Inherently, cassava is able to survive longer in drought areas. During periods of moisture deficit the cassava plants lose all their leaves thereby suspending production of photosynthates and translocation to the bulking roots. When moisture is restored, the plants regenerate without much loss in yield occurring. This ability of cassava to survive in arid conditions gives it an advantage as a crop of choice in extremes of weather and climate in the southern Africa region.

Genetically, uniform modern varieties must not replace the highly diverse local cultivars and landraces in traditional agroecosystems but complement them. Uniformity may entail a disaster when extreme weather occurs that is detrimental to the uniform variety. Therefore, breeding must be of necessity mimic and keep the diversity of local cultivars in traditional farming systems but improve the productivity of the varieties besides generating highly productive materials for commercial use.

4.5 FARMER INVOLVEMENT IN CLIMATE SMART TRAITS EVALUATION

Participatory varietal selection (PVS) is a tool widely used by breeders albeit its limitations and complexity. The development of PVS techniques originated in the social sciences division of agriculture and the current use by breeders has not fully utilized the data generated and connected to materials, which have made impact in the seed market. The strength of using PVS lies in being consultative and participatory nature, which appeals to the majority of stakeholders. In most countries, PVS data in particular preference votes of farmer groups in support of a breeding line is a prerequisite for a line to be considered for the official release by the seed certifying authorities.

In most cases, the participatory approach is not followed or applied correctly, otherwise some progress has been made and more can be achieved through this approach. It needs a multidisciplinary team to work together for better results. Constituting multidisciplinary teams requires adequate resources, which are often limiting.

4.6 MAKING BREEDING PRODUCTS AVAILABLE ON CLIMATE AFFECTED FARMS

Climate adaptive crop cultivars are the building blocks to improved cropping systems. To reduce the impact of extremes of weather, crop cultivars that are drought resistant, heat, and cold tolerant need to be disseminated to farm families as a matter of urgency. The cultivars also need to tolerate weather-induced pests and disease outbreaks that frequently occur and can cause enormous damage to crops.

4.6.1 Regulatory Framework on Seed Release

The regulatory framework for seed lease has a huge bearing on the access of breeding materials by farmers in each country in southern Africa. In general, each country has a seed release system in place but the efficiency with which they operate vary widely. Two major components of the seed regulation procedure, which affect seed release to the public are: (1) the period required to evaluate and determine the characteristic data for a cultivar and (2) the practical arrangements of managing the seed after the release process.

Seed certification takes place after (distinctiveness, uniformity, stability) DUS has been verified by the authority. The amount of time it takes could range from one to three growing season. Coupled with the 7 years or so required to develop a variety, this brings the total to 10 years before a farmer can benefit from climate adaptive seed.

When a variety is released, the breeder responsible over the variety has the sole mandate to maintain the seed and facilitate its presence in the market. The promotion pathway for a released variety in most cases is not very well articulated in the release procedures. Seldom do released materials immediately get publicity and hence farmers can use them within a short period of time.

The current regulation in summary allows for a register of new varieties into the market; however, no consideration has been given to delist (from the register) varieties that have broken down leading to production problems at farm level. There is need for reforms that favor promotion of modern varieties that can overcome climate change challenges and delisting of those cultivars that are prone to a range of constraints to include pests and diseases.

4.7 CONCLUSION

Over and above higher yields, breeding can build drought, cold and heat tolerance, and earliness feature, which are critical for climate smart agricultural development. In addition to conventional breeding, new molecular and genomic tools can be employed to speed up the incorporation of desirable genes for the development of future climate smart resilient varieties.

Many new improved varieties are environmentally friendly, ensuring food security, while conserving the environment. Genetic diversity and plant breeding are key elements in tackling climate change, and integration of plant breeding in climate change strategies is one of the best paths to sustainable food production. Environmentally friendly varieties are improved varieties resistant to pests, which require fewer pesticides. High-yielding varieties are those that enable increased food production per unit area and alleviate pressure to add more arable land to production systems.

Bureaucracy in variety release is one of the factors that hinder the adoption of climate resilient improved varieties. When it is not officially realized, it is not possible to have mass multiplication of these varieties, which have been identified and selected at on-farm level denying farmers the much needed varieties. Low seed multiplication ratio in crops such as cassava necessitates mass seed production action after being selected. It is a must to realize this. Therefore, reduced bureaucracies are essential for delivering climate smart breeding products to small-scale farmers.

REFERENCES

Arega, A.D., Khataza, R., Chibwana, C., Ntawuruhunga, P., Moyo, C., 2013. Economic impacts of cassava research and extension in Malawi and Zambia. Journal of Development and Agricultural Economics 5 (11), 457–469.

Benesi, I., Minde, I., Nyondo, F., Trail, T., 1995. Accelerated Multiplication and Distribution of Cassava and Sweet Potato Planting Materials as a Drought Recovery Measure in Malawi. Adoption Rate and Impact Assessment Study. USAID. http://pdf.usaid.gov/pdf_docs/PNABZ355.pdf.

Bolanos, J.S., Edmeades, G.O, 1996. The importance of the anthesis-silking interval in breeding for drought tolerance in maize. Field Crops Research 48, 65–80.

Carena, M.J., 2005. Registration of NDSAB(MER-FS)C13 maize germplasm. Crop Science. 45, 1670–1671. http://www.southernafricatrust.org/wp-content/uploads/2014/09/can-smallholder-farmers-address-hunger-in-the-region.pdf.

Ceballos, H., Kawuki, R.S., Gracen, V.E., Yencho, G.C., Hershey, C.H., 2015. Conventional breeding, marker-assisted selection, genomic selection and inbreeding in clonally propagated crops: a case study for cassava. Theoretical Applied Genetics. http://dx.doi.org/10.1007/s00122-015-2555. https://www.researchgate.net/publication/231870954_Plant_breeding_and_climate_changes_J_Agric_Sci.

Crush, J., Frayne, B., 2010. The Invisible Crisis: Urban Food Security in Southern Africa. AFSUN, Cape Town.

Fehr, W.R., 1984. Genetic Contributions to Yield Gains of Five Major Crop Plants. Crop Science Society of America and American Society of Agronomy.

Haggblade, S., Nyembe, M., 2008. Commercial dynamics in Zambia's cassava value chains. Food Security Research Project, Lusaka, Zambia.

Kotir, J.H., 2011. Climate change and variability in Sub-Saharan Africa: a review of current and future trends and impacts on agriculture and food security. Environment, Development and Sustainability 13 (3), 587–605.

Legg, J.L., Kumar, P.L., Makeshkumar, T., Tripathi, L., Ferguson, M., Kanju, E., Ntawuruhunga, P., Cuellar, W., 2015. Cassava virus diseases: biology, epidemiology, and management. In: Loebenstein, G., Katis, N.I. (Eds.). Loebenstein, G., Katis, N.I. (Eds.), Advances in Virus Research, 91. Academic Press, Burlington, pp. 85–142.

Lobell, D.B., Burke, M.B., Tebaldi, C., Mastrandrea, M.D., Falcon, W.P., Naylor, R.L., 2008. Prioritizing climate change adaptation needs for food security in 2030. Science 319 (5863), 607–610.

Mackay, I., Horwell, A., Garner, J., White, J., McKee, J., Philpott, H., 2011. Reanalyses of the historical series of UK variety trials to quantify the contributions of genetic and environmental factors to trends and variability in yield over time. Theoretical and Applied Genetics 122 (1), 225–238.

Ntawuruhunga, P., Andrade, M., Demo, P., Moyo, C.C., 2011. Improving the rural livelihoods in Southern Africa project: 2004–2010. Roots 13, 14–18.

Rowntree, S.C., Suhre, J.J., Weidenbenner, N.H., Wilson, E.W., Davis, V.M., Naeve, S.L., Casteel, S.N., Diers, B.W., Esker, P.D., Specht, J.E., Conley, S.P., 2013. Genetic gain × management interactions in soybean: I. Planting date. Crop Science 53 (3), 1128–1138.

Tefera, H., Asafo-Adjei, B., Dashiell, K.E., 2010. Breeding progress for grain yield and associated traits in medium and late maturing promiscuous soybeans in Nigeria. Euphytica 175 (2), 251–260.

CHAPTER 5

Enhancing Gains From Beneficial Rhizomicrobial Symbiotic Communities in Smallholder Cropping Systems

Nhamo Nhamo[1], George Mahuku[2], David Chikoye[1], John O. Omondi[3]

[1]International Institute of Tropical Agriculture (IITA), Southern Africa Research and Administration Hub (SARAH) Campus, Lusaka, Zambia; [2]International Institute of Tropical Agriculture (IITA), East Africa, Dar es Salaam, Tanzania; [3]Ben Gurion University of the Negev, Beer Sheba, Israel

Contents

Smart Technologies for Sustainable Smallholder Agriculture
ISBN 978-0-12-810521-4
http://dx.doi.org/10.1016/B978-0-12-810521-4.00005-0

5.1 INTRODUCTION

Climate change will affect the below-ground biodiversity, the functionality of collaborative microbial communities in cropping systems, and the processes driving nutrient cycling in major soil groups. The fauna communities associated with agroecology will be affected the most. The whole range of micro-, meso-, and macrofauna and the related food webs will shift in response to both rainfall and temperature variability as niche conditions change. IPCC (2007) has projected shifts in species composition in response to climate change. These shifts will affect the ecology and interactions of microbial communities with symbiotic relations in the rhizosphere. The rhizosphere dynamics has large influence over water, plant nutrients, pathogens, and beneficial relations on which cropping systems depend on for adapting to environmental changes. Climate variability in Southern Africa has a direct bearing on the soil moisture (mainly a function of precipitation) and niche characteristics (influenced by organic matter, temperature, moisture, and nutrient changes). Both species composition and diversity may simultaneously change because of climate variability.

Soil biology research has evolved over time and significant improvements have been in the thematic area of use of resident microorganisms and external inoculants to enhance crop growth and reduce crop pests and disease.

5.2 DEFINING BENEFICIAL SYMBIONTS FOR NITROGEN FIXATION, CROP ENHANCEMENT, AND CROP PROTECTION

Soil is rich in microorganisms that are both beneficial and detrimental to crop production. Estimates show that only a third of this wealth of microbes has been explored and an even small proportion (>1%) has been studied and characterized to depth. These estimates show the potential of soil inhabiting organisms in the development of future solutions to ecological challenges the planet faces. Since the early years of agriculture, bacterial and fungal genera have been studied for beneficial effects on agriculture. To date, beneficial isolates of both bacterial and fungal genera are widely used as cultured inoculants and indigenous populations that can enhance crop production. Notable developments are the use of *Rhizobia* species in the biological nitrogen fixation (BNF) in the production of legumes, arbuscular mycorrhizae (AM) to enhance crop growth, and nonaflatoxigenic associations (NAAs) to prevent crop infections.

5.2.1 Biological Nitrogen Fixation

The study of rhizosphere biological interaction dates as far back as the 17th century when bacterial species were discovered to be in close association with plant roots. Successful isolation of these microorganisms occurred only in the 19th century, and the genera of importance then consisted of *Bradyrhizobium*, *Sinorhizobium*, *Azorhizobium*, *Mesorhizobium*, and a few others. These were the dominant constituents of the genus *Rhizobium* (rhiza = root; bios = life). Nodules formed on the roots of legumes contain symbiotic bacteria that are essential for BNF, which in turn contribute to the nitrogen budget in the soils of farmlands. The discovery of the term biological nitrogen fixation is attributable to two scientists, i.e., Hermann Hellriegel and Martinus Beijerinck, and it is the process by which atmospheric nitrogen (N_2) is converted to ammonia (NH_3), catalyzed by the enzyme nitrogenase. During the process, adenosine triphosphate (ATP) molecules are hydrolyzed, hydrogen molecules are formed, protonation of nitrogen molecules occurs, and reduction process proceeds on active iron-molybdenum cofactor sites Eq. (5.1). In the modern application of legumes in cropping systems, rhizobial inoculation has become a beneficial practice, which enhances the growth of and resource use by legumes. Several successful studies have been reported to include the wide use and support through infrastructure development, e.g., in Malawi, South Africa, Zambia, and Zimbabwe where inoculant products are widely sold.

$$N \equiv N + 8H^+ + 8e^- + 16ATP \rightarrow 2NH_3 + 16ADP + H_2 + 16Pi \quad (5.1)$$

Although the contribution of legumes to human and animal diets and BNF in faming systems are widely recognized, current trends have not translated into more investment in production of these crops. More research work is still required to get fixation rates that are higher, more efficient, and can demonstrate superiority over mineral fertilizers. This is more pronounced given the impending challenges with climate change and the cost of energy, which directly translate into higher cost of production of fertilizer inputs. In southern Africa, BNF using suitable legumes has huge potential to enhance production of legumes and associated crop (food, nutrition security, and generation of income) and mitigate climate change by reducing dependency on mineral fertilizer sources (fertilizers are costly, increase emissions, and consume a lot of fuel power).

5.2.2 Root–Fungus Associations

A parallel research development took place with emphasis on the manipulation of vesicular arbuscular mycorrhizal fungi (AMF) to enhance crop growth of host plants. Mycorrhiza (*mykos* = fungus; *riza* = root) is a symbiotic association between fungi and roots of a vascular crop. Mycorrhization is the relationship between the host plant and the fungus involved in both colonization and infection. Although research started in the 1880s, isolation and identification were done between 1952 and 1957. The three structures that have been used in characterizing mycorrhiza are (1) the hyphae, (2) arbuscules, and (3) vesicles. The vesicles are prejudicial and suggest a host plant being diseased by fungi. Although mycorrhizal association is considered mutual to both crops and fungi, there are reports that the relationship can turn out to be weakly pathogenic, i.e., the crop ends up being parasitized by the fungi. Negative effects on plants have been reported in which the mycorrhiza occurs in nonhost plants and the soil phosphorus availability is reduced.

5.2.2.1 Fungal Infections

Two mycorrhizal associations dominate major crops and trees: (1) AM, formerly vesicular AM, and (2) sheathing mycorrhizae, e.g., ectomycorrhiza, arbutoid mycorrhiza, and monotropoid mycorrhiza. Colonization in AM is through intracellular colonization by aseptate obligate fungi belonging to the order Glomales (Glomeromycota). The sheathing is formed by the septate fungi from the families Ascomycota and Basidiomycota, and the colonization is intracellular.

In AMF, these structures are important: arbuscules, coils, vesicles, intraradial mycelium, and extraradial mycelium. The mycelium explores the soil for nutrients and is useful in the transportation of nutrients from the soil.

By growing mycorrhiza in cropping systems the plant benefits from the increased absorption of water and nutrients by the mycelia, which have higher absorptive capacity and explore a large soil surface area. In return the association supplies the fungal partners with carbohydrate sugars, including glucose and sucrose.

On the other hand, mycorrhizal association will provide the required solution to water stress, nutrient acquisition, and protection from pests and diseases.

Climate-smart agriculture targets to improve water use efficiency of crops grown in drought-prone areas of Southern Africa and the level of contribution of mycorrhizal association to this effect is uncharacterized for most food crops and nutritious legumes.

There is scope in emphasizing the cultural practices and use of micro-symbionts as a part of the crop intensification and good agricultural practices. However, careless application of agrochemicals, e.g., herbicides, can reduce the prevalence of mycorrhizal fungal communities in the soil, leading to problems.

5.2.2.2 Commercial Products of Mycorrhizal Fungi

Contrary to rhizobia, mycorrhizal fungi inoculants have not been popularized in Southern Africa. However, inoculants for mycorrhizae are widely distributed and used in the United States, e.g., Myconate. To date, most of the AM are a result of indigenous fungal populations that infect crop roots. The risk of limiting the potential of growing these associations in many soils is high because of practices such as use of fire for clearing land and excessive agrochemicals for pests (insects and weeds), and hence the need for the development of AM inoculants that support MA growth on agricultural crops. More work is required in this area, given the benefits and the fit for such a technology in Southern Africa.

5.2.3 Nonaflatoxigenic Plant Associations

5.2.3.1 Aflatoxin-Producing Fungi

Mycotoxins are low-molecular-weight secondary metabolites produced by certain strains of filamentous fungi, which if ingested induce various degrees of toxicity to vertebrates, invertebrates, plants, and microorganisms. Of the numerous mycotoxins, aflatoxins are produced predominantly by certain strains of the genus *Aspergillus* and are the most widely studied. They have immunosuppressive, mutagenic, teratogenic, and carcinogenic effects, especially on the liver (Wu et al., 2012; Wu and Khlangwiset, 2010). Aflatoxin contamination of maize, the major cereal in African diets, is a major risk for the health and well-being of African people, primarily children. When consumed in low dosages over prolonged periods, aflatoxins may cause liver cancer, suppress immune systems, increase the incidence and severity of infectious diseases, and retard child growth and development by contributing to malnutrition (Gong et al., 2008, 2004; Cardwell and Henry, 2004; Turner et al., 2003). Young animals and children are the most sensitive to the effects of consuming aflatoxin-contaminated food (Williams et al., 2004). The most important producers of aflatoxin are *Aspergillus flavus* and *Aspergillus parasiticus*, and these species are ubiquitous in many tropical soils where maize and groundnuts are grown. Grains can be infected from preharvest stages in the field to postharvest stages during storage.

5.2.3.2 History of Aflatoxin Research

Aflatoxins were discovered in the early 1960s when they were identified as causative agents of "turkey X" disease, an epidemic involving the death of numerous turkey poults, ducklings, and chicks fed diets containing peanut meal imported from Brazil (Blount, 1961). The turkey X disease killed approximately 100,000 birds near London in England (Blount, 1961; Forgacs and Carll, 1962). Investigations revealed that toxicity was associated with the presence of *A. flavus*, and extracts from the cultures of the fungus isolated from the meal were capable of inducing the turkey X disease. The name aflatoxin (*A. flavus* toxin) was accordingly assigned to the toxic agents.

These findings stimulated extensive research efforts to assess potential health hazards resulting from contamination of the human food supply chain and to minimize exposure. Findings that aflatoxins were carcinogenic caused concern over their occurrence in human foods and led to worldwide efforts to prevent their occurrence in human food and animal feed and to determine the relationship these carcinogens with human diseases. The findings that aflatoxins were immunosuppressive and that they were probably the underlying cause of other diseases stimulated research in aflatoxins. Subsequent efforts led to the finding that aflatoxins can occur before harvest and therefore they were no longer only a storage problem. Major crops such as maize, groundnuts, cottonseed, and certain tree nuts are frequently found to be contaminated with aflatoxins. These findings resulted in a large multidisciplinary scientific investigation on the various aspects of concern such as eradication, control, analysis, epidemiology, and plant pathology as well as major efforts to determine the nature of the human and animal diseases they cause. Present-day investigations with aflatoxins continue with elimination as a major thrust based on knowledge of the biosynthetic pathway, genetics of both host and pathogen, host-parasite-vector interactions, plant breeding, biocontrol, and selected agronomic practices. We will focus on advances made to date to develop and use atoxigenic strains of *A. flavus* and to minimize aflatoxin contamination of major commodities.

5.2.3.3 Chemical Composition of Aflatoxins

Four types of aflatoxins occur naturally: B1, B2, G1, and G2. These toxins are differentiated based on their fluorescence under ultraviolet light (Agag, 2004). Members of the blue fluorescent (B) series are characterized by the fusion of a cyclopentenone ring and lactone ring of the coumarin moiety, whereas the green fluorescent (G) toxins contain a fused lactone ring. Aflatoxins B1 and B2 (AFB1 and AFB2, respectively) were so named because of their strong blue fluorescence in ultraviolet light, whereas aflatoxins G1 and G2 (AFG1 and AFG2, respectively) fluoresced greenish yellow. In the early 1960s, these properties facilitated the rapid development of methods for monitoring grains and other food commodities for the presence of the toxins. Of these aflatoxin forms, AFB1 is considered the most toxic (Boonen et al., 2012).

5.2.3.4 Drivers of Aflatoxin Contamination in Southern Africa

Biotic and abiotic factors, either nutritional or environmental, are known to affect aflatoxin production in toxigenic *Aspergillus* species, but the molecular mechanisms for these effects are still unclear (Payne and Brown, 1998). Nutritional factors such as the presence of carbon, nitrogen, amino acids, lipid, and trace elements affect aflatoxin production (Feng and Leonard, 1998; Payne and Brown, 1998; Cuero et al., 2003). Temperature, pH, water activity (drought stress), and other stresses are external environmental factors that can affect aflatoxin production (Cotty, 1989; Payne and Brown, 1998). Optimal aflatoxin production is observed at temperatures near 30°C (28–35°C). High temperature and drought, which often occur together during the growing season and likely contribute to poor kernel development, have been reported to increase growth of the fungus and toxin production (Payne, 1998). Irrigation not only relieved drought stress but also reduced soil temperature. Increased aflatoxin contamination was observed in drought-treated peanuts with increased soil temperatures (Cole et al., 1985). Mc Millian et al. (1985) conducted a 6-year study in which the 3 years with the highest contamination also had the highest average daily temperatures during the growing season. Similarly, in a 5-year study, a significant positive correlation between aflatoxin contamination and temperatures was obtained only during the 2 years with exceptionally high concentrations of aflatoxin (Widstrom et al., 1990).

5.2.3.5 Exposure to Aflatoxins

Humans are exposed to aflatoxins following consumption of contaminated grains and animal products. Although aflatoxin contamination of commodities may be universal, the levels or final concentrations of aflatoxins in the grain product can vary from less than 1000 μg/kg (1 ppb) to greater than 12,000 μg/kg (12 ppm). In an outbreak of aflatoxin-induced death of people in Kenya, the individual daily exposure of AFB1 was estimated to be 50 mg/day (Probst et al., 2007). When AFB1 and AFB2 are hydroxylated in animal tissues such as milk, meat, and eggs, aflatoxin M1 and M2 metabolites are produced, respectively (CAST, 2003; Agag, 2004; Richard, 2007). Therefore, minimizing the contamination of grains with aflatoxins in the field (preharvest) is the most economical way to minimize human exposure.

5.2.3.6 Examples of Aflatoxin Poisoning
In the 1970s, aflatoxin poisoning occurred in western India, leading to at least 97 fatalities. These deaths occurred only in households where heavily contaminated maize had been consumed. Histopathology of liver specimens revealed extensive bile duct proliferation, a lesion often noted in experimental animals after acute aflatoxin exposure (Bhat and Krishnamachari, 1977; Krishnamachari et al., 1975). In early 1980 an incident of acute aflatoxicosis was reported in Kenya, where 20 hospital admissions resulted in 20% mortality and this was associated with consumption of maize highly contaminated with aflatoxin (Ngindu et al., 1982). Consecutive outbreaks of acute aflatoxicosis in Kenya in 2004 and 2005 caused more than 150 deaths. In April 2004, one of the largest documented aflatoxicosis outbreaks occurred in rural Kenya, resulting in 317 cases and 125 deaths (Azziz-Baumgartner et al., 2005; Lewis et al., 2005). Aflatoxin-contaminated maize grown and eaten on family farms was the major source of the outbreak. In a survey of 65 markets and 243 maize vendors, 350 maize products were collected from the most affected districts. About 55% of maize products had aflatoxin levels greater than the Kenyan regulatory limit of 20 ppb, 35% had levels >100 ppb, and 7% had levels >1000 ppb. In addition to the market survey for aflatoxin exposure, this outbreak marked the first time that biomarkers, namely, aflatoxin–albumin adducts, were used to independently confirm the exposure in individuals (Azziz-Baumgartner et al., 2005; Lewis et al., 2005; Probst et al., 2007; Strosnider et al., 2006).

5.2.3.7 Breakthroughs in Aflatoxin Research: Development, Validation, and Application of Aflatoxin Biomarkers
A biomarker of exposure refers to the measurement of the specific agent of interest, its metabolites, or its specific interactive products in a body

compartment or fluid, which indicates the presence and magnitude of current and past exposure. Several types of biomarkers have been developed and have revolutionized the assessment of exposure as a function of dietary diversity. Urinary measures of aflatoxin M1, aflatoxin-mercapturic acid, and the aflatoxin–albumin adduct have been shown to correlate with aflatoxin exposure (Groopman et al., 1992a,b; Wild et al., 1992). Analytical methods have been developed for the quantitation of aflatoxin metabolites, aflatoxin–DNA adducts, and aflatoxin–serum albumin adducts in biological samples (Poirier et al., 2000; Santella, 1999; Gan et al., 1988; Groopman et al., 1985). These advances have allowed research to identify and document the link between aflatoxins and certain diseases, such as child stunting, immunosuppression, and incidence of cancers. In addition, these advances have allowed research to look at interventions to minimize aflatoxin exposure, such as the use of clays with high affinity for aflatoxins, which bind aflatoxin in the human gut and prevent its uptake.

5.2.3.8 Current Interventions

The goal and hence the strategy used is prevention by reducing exposure to aflatoxins in the diet of communities in high-exposure locations. A range of interventions are available, including planting pest-resistant varieties of staple crops, attempting to lower mold growth in harvested crops, improving storage methods following harvest, and using trapping agents that block the uptake of unavoidably ingested aflatoxins. Many strategies, including biological control, control of insect pest, and development of resistant cultivar, have been investigated to manage aflatoxins in crops. Among them, biological control appears to be the most promising approach for control of aflatoxin in both preharvest and postharvest crops. The use of nontoxigenic *Aspergillus* fungi as a strategy to minimize pre- and postharvest aflatoxin contamination has received greatest attention and has consistently reduced contamination by more than 80% (Abbas et al., 2011). Research has shown that aflatoxin production decreases in maize inoculated with nontoxigenic *A. flavus* strains before inoculation with toxigenic strains (Brown et al., 1991). The potential for biological control of aflatoxins through competitive exclusion has been demonstrated under field conditions in cotton (Cotty, 1994, 2006), groundnuts (Dorner, 2009), and maize (Abbas et al., 2006, 2009). To date, the most effective strategy for reducing aflatoxin levels in harvested corn is biological control using nonaflatoxigenic strains of *A. flavus*.

5.2.3.9 Criteria for the Selection of a Biological Control Agent

Aspergillus is ubiquitous in tropical and temperate soils. Both toxigenic and nontoxigenic *A. flavus* strains coexist in all these environments, and the ability of nontoxigenic *A. flavus* strains to compete effectively for the same ecological niche provides the basis for biological control. Identification of atoxigenic strains suitable for use in biocontrol involves several approaches, including the use of phylogenetics and phenotypes that may or may not reflect phylogeny. The most useful characteristic would be one that correlates well with non-production of the major aflatoxins (B1, B2, G1, G2). Phenotypically, *A. flavus* strains can be categorized on the basis of sclerotium size. Small sclerotia (<400 μm in diameter) are associated with toxin production, whereas strains producing large sclerotia (>400 μm) may be toxigenic or nontoxigenic (Abbas et al., 2005, 2009; Cotty, 1997; Horn, 2003). Vegetative compatibility groups (VCGs) have been used in many surveys, but the technique is labor intensive and the results of one study are often not readily compared to other studies. Also there is a tremendous amount of diversity within VCGs of *A. flavus* populations. For example, a survey of 255 soil isolates of *A. flavus* from one field identified 16 VCGs (Sweany et al., 2011).

The nontoxin-producing trait is a useful marker to isolate a potential afla-toxin biocontrol strain (Abbas et al., 2011). Additionally, even if toxins are not detected in culture, the presence of toxin biosynthetic genes should signal the need for further scrutiny before large-scale application of the agent. Therefore, sequencing the aflatoxin biosynthetic gene pathway is essential to make sure that isolates selected as potential biocontrol have lost the ability to produce aflatoxins. Following this confirmation, VCG analysis is conducted to make sure that the potentially atoxigenic isolates do not belong to a VCG that also contains toxin producers. Isolates in the same VCG can potentially exchange genetic material. The biological control strains that do not produce aflatoxins in culture are nonfluorescent on 0.3% β-cyclodextrin media when exposed to ultraviolet light (365 nm) and do not produce yellow pigment on potato dextrose agar (PDA) medium, and this characteristic can be used to check when selecting potentially atoxigenic strains for use as biocontrol agents. Molecular approaches based on DNA sequences are increasingly being used, as they are relatively cheap, are fast, and can reveal functional and phylogenetic relationships between isolates. Some recently applied DNA-based approaches include pyrosequencing (Das et al., 2008) and a multitude of PCR-based approaches (Chang et al., 2005; Jiang et al., 2009; Niessen, 2007) to make sure that atoxigenic strains selected as potential biocontrol products are phyloge-netically distant from toxin producers.

5.2.3.10 Efficacy of Biocontrol Strains

The primary goal of developing biocontrol strategies is to minimize myco-toxin contamination in crops, specifically maize and groundnuts. Atoxigenic strains are applied to the field during crop development (Cotty, 1994; Dorner and Lamb, 2006). Under appropriate conditions, the spread of the introduced strain throughout the field displaces the native, toxic strains (Mehl et al., 2012; Atehnkeng et al., 2014). Strains formulated into biologi-cal control products may be single clones (Bock and Cotty, 1999) or com-posed of more than one strain to improve local adaptability (Atehnkeng et al., 2014). Several climatic factors have been identified that affect efficacy. Dew and moisture will allow for the atoxigenic strains to produce spores over several days. Timing of biocontrol application is crucial for success. Late application of atoxigenic strains on maize (after silking) may not be effective. Application of biological control has consistently been shown to reduce aflatoxin contamination by more than 80% and the effect carries into storage. Several methods can be used to deliver biocontrol products, but the most effective is one that does not demand much from the farmer's time. Soil application of biocontrol *A. flavus* strains is effective at reducing aflatoxins in specific harvested crops, but direct application to maize ears is more effective. For example, Lyn et al. (2009) reported that spraying non-toxigenic *A. flavus* directly on corn silks reduced aflatoxin contamination by 97%, whereas soil application of the same strain coated on sterile sorghum seeds reduced aflatoxin by 65%. However, spraying maize silks is labor intensive and will not be most likely adopted by smallholder farmers in Africa. Studies conducted by the International Institute of Tropical Agriculture (IITA) have revealed that delivery of atoxigenic strains immo-bilized of sterile sorghum is a very effective delivery mechanism.

5.2.3.11 Mechanism of Control

Biological control measures to mitigate aflatoxin contamination are consid-ered as a competitive interaction of toxigenic and nontoxigenic fungal strains (Dorner, 2004). This competition can be divided into two basic types: *interference* and *exploitative* competition. Interference competition includes antibiosis, such as production of inhibitory surfactins (Mohammadipour et al., 2009), production of extracellular hydrolytic enzymes by *Trichoderma* species (Gachomo and Kotchoni, 2008), and production of antifungal anti-biotic pyrrocidines A and B by the maize endophyte, *Acremonium zeae* (Wicklow et al., 2005). Exploitative competition, in contrast, is simply the occupation of an ecological niche, thus depriving the competitor of space and essential nutrients. Exploitative competition is almost certainly a major element in the biological control of aflatoxin-producing *A. flavus* strains by nonaflatoxigenic *A. flavus*. However, more research is required to better understand the mechanism of biocontrol with atoxigenic strains under different climatic conditions.

5.2.3.12 Use Under African Conditions

Use of atoxigenic *A. flavus* strains to mitigate aflatoxin contamination in staple foods has been adopted in Africa (Atehnkeng et al., 2014). In one study in Nigeria, the inoculation of a mixture of four endemic atoxigenic strains of *A. flavus* in maize plots in four agroecologies over 2 years resulted in significant reductions in aflatoxin concentrations at harvest and after storage (Atehnkeng et al., 2014). At harvest, the reduction in aflatoxin levels ranged from 57.2% (27.1 ppb in untreated plots vs. 11.6 ppb in treated plots) to 99.2% (2792.4 ppb in untreated plots vs. 23.4 ppb in treated plots). The applied atoxigenic strains remained in the treated crops, and the reduction in aflatoxin concentrations in grains after poor storage ranged from 93.5% (956.1 ppb in untreated vs. 66.2 ppb in treated) to 95.6% (2408.3 ppb in untreated vs. 104.7 ppb in treated).

In Nigeria, a similar percentage of maize samples were contaminated by both aflatoxin and fumonisin (Adetuniji et al., 2014), which is not uncommon. When conditions are permissive for both aflatoxin and fumonisin production in the field, interventions that are effective for both toxins are needed. Aside from Bt maize, which is not yet widely used in Africa, there are few interventions for preemptive prevention of fumonisin in the field. Preliminary trials have shown potential for the development of biological control treatment for *Fusarium verticillioides* (Sobowale et al., 2007).

Genetic recombination in *A. flavus* has been shown to increase genetic variation within the population (Olarte et al., 2012). Sexual recombination leading to the acquisition of toxin genes is possible, but the implications of this are not clear with respect to biological control (Abbas et al., 2011). Studies to date show that aflatoxin production is heritable and is not lost during sexual recombination; however, hybridization between toxic and atoxigenic strains produced progeny with no or lower aflatoxin production (Olarte et al., 2012).

5.2.3.13 Effect of Aflatoxins on Human Health

Aflatoxins are chemical compounds that are produced by a range of fungal species from the *Aspergillus* genus, which are potential carcinogens and teratogens to humans and farm animals (Cotty et al., 2007). Spores of these fungi are common in air and soil of agricultural areas of temperate and tropical environments. A variety of species of the fungal genus *Aspergillus* (mainly *A. flavus* and *A. parasiticus*) synthesize aflatoxins.

Aflatoxins are highly toxic, cancer-causing chemicals produced by several fungal species within *Aspergillus* section Flavi. Presence of aflatoxins in human foods causes acute and chronic health effects (aflatoxicosis), ranging from immune system suppression, growth retardation, and cancer to death from acute poisoning (Wild and Turner, 2002).

In developed countries, stringent government regulations limit the use of aflatoxin-contaminated crops in foods and feeds, and hence, commodities with aflatoxin content exceeding the maximum permissible level have significantly diminished cash value.

5.2.4 Legumes and Rhizobial Technology

The technology of inoculating grain legumes with rhizobia to enhance BNF has been widely explored (Abaidoo et al., 2007; Maingi et al., 2006; Mulas et al., 2015; Omondi et al., 2014). The four principal factors guiding BNF are enumerated by Van Kessel and Hartley (2000) as the effectiveness of rhizobia–host plant symbiosis, the ability of the host plant to accumulate N_2, the amount of available soil N_2, and the environmental constraint to N_2 fixation. However, of these four factors, the first one has been extensively studied (Fening and Danso, 2002; Mulas et al., 2015) leading to the development of inocula that are used in BNF. Benefits of these inocula are observed as an increase in both legume grain yields and the quality and enhancement of soil fertility through the N_2 fixed (Bloem et al., 2009; Mafongoya et al., 2006). These benefits have spread from the point of innovation to other areas growing legumes, including Southern Africa. Fig. 5.1 indicates production and sales of inoculants in Zimbabwe, a Southern African country (Giller et al., 2000). Moreover, a review on the production and adoption of inoculants in Southern Africa depicts growth (Bala et al., 2011). This uptake in Southern Africa notwithstanding, Abate et al. (2012) report that legume yield improvement is still facing myriad of problems and the yields are below world averages. However, they attribute increase in legume production in sub-Saharan Africa to increase in area of production rather than yield per area. This fact of area of production expansion is further elucidated by Rachel et al. (2007) in their participatory research in Malawi. Thus this necessitates more studies and reviews of the remaining three factors proposed by (Van Kessel and

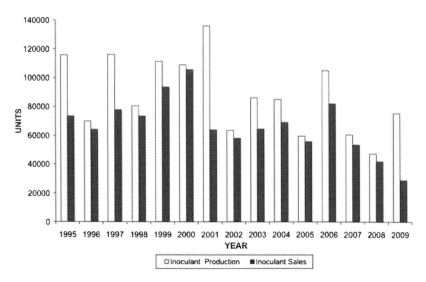

Figure 5.1 Production of soybean inoculants (indicated in number of 100-g sachets) at SPRL, Marondera, Zimbabwe, since 1995. *Giller, K.E., Murwira, M.S., Dhliwayo, D.K.C., Mafongoya, P.L., Mpepereki, S., 2000. Soyabeans and sustainable agriculture Promiscuous soyabeans in Southern Africa. Fields Crop Research 65, 137–149.* http://dx.doi.org/10.1016/S0378-4290(99)00083-0.

Hartley, 2000) to achieve maximum BNF and grain yield in Southern Africa.

In their study, Indieka and Odee (2005) observed soil N_2 and environment influencing BNF as the main factor to consider, which indeed applies to Southern Africa (Bloem et al., 2009; Cramer et al., 2010; Lemaire et al., 2015). However, in exploring this phenomenon, coupling with specific agroecological zones and rhizobia specificity has remained obscure and underdiscussed. Three main factors stand out as the major determinants of high nitrogen fixation: (1) the legume species involved, (2) the rhizobia strain involved, and (3) the effectiveness of the symbiosis in converting atmospheric nitrogen to ammonia. These three factors together with climatic factors will determine the overall gain BNF provides from the N cycles to smallholder farms. Giller (2001) summarized the N_2 yields for the four legumes as shown in Table 5.1.

Furthermore, legumes respond favorably to other management practices and the input available at the farm, for instance, groundnuts, responds favorably to the application of gypsum, resulting in increased number of pegs and hence grain yield.

Table 5.1 Amount of nitrogen fixed by various legumes: a contribution to the nitrogen budget on smallholder farms

Legume	N$_2$ Fixed (kg/ha)	Source
Bambara nut	52	Rowe and Giller (2003)
Cowpea	47	Rowe and Giller (2003)
Groundnut	33	Rowe and Giller (2003)
Pigeon pea	39	Rowe and Giller (2003)
Pigeon pea	3–82	Mapfumo et al. (2000)
Pigeon pea	97	Chikowo et al. (2004)
Cowpea	28	Chikowo et al. (2004)
Acacia angustissima	122	Chikowo et al. (2004)
Sesbania sesban	84	Chikowo et al. (2004)
Gliricidia sepium	212	Mafongoya PL
A. angustissima	210	Mafongoya PL
Leucaena collinsii	300	Mafongoya PL
Tephrosia candida	280	Mafongoya PL
Tephrosia vogelii	157	Mafongoya PL

Source: Mafongoya, P.L., Bationo, A., Kihara, J., Waswa, B.S., 2006. Appropriate technologies to replenish soil fertility in Southern Africa. Nutrient Cycling in Agroecosystems 76, 137–151. http://dx.doi.org/10.1007/s10705-006-9049-3.

5.2.5 Increasing Benefits and Scale of the Legume Technology

The major benefit of applying rhizobia is the increase in both biomass and grain yield. These two are important for human consumption (grain), soil fertility amendments, and livestock feed (biomass and grain in feed formulations). Grain yield has steadily increased from an average of about 500 kg/ha to about 1500 kg/ha. These results have been attributed to introduction of improved cultivars that are higher yielding. In the recent past, breeding for improved yields has not further increased the yields significantly. This suggests that some of the cultural practices that came with the use of inoculant and other inputs on crops could have influenced the general care and maintenance farmers applied on the crops. For instance, the application of inoculant encourages farmers to (1) plant on time with the right amount of moisture, (2) observe the right amount of seeding rates per unit area, (3) practice weeding to ensure higher yields, (4) scout for any disease incidence, and (5) apply P fertilizer as a requirement. We hypothesize that a large proportion of the yield gap reduction also comes from the combined effect of good management practices farmers pick up for legume production relative to the various characteristics.

Scale requires improved sensitization to socioeconomic factors leading to reduced production. These include household perceptions on the value of legumes that influence the decision to limit allocated land, input directed on legumes, labor devoted to legume plots, and the marketing effort for legumes. One major driver of production is organized markets, and in the case of soybeans, the involvement of private sector has led to improved production and value from the crop. The expansion of legume value chains beyond local markets and from few kilograms to numerous tonnes of sale volumes will greatly improve the visibility of legumes.

5.3 HARNESSING MYCORRHIZAL BENEFITS IN DEGRADED SOILS

Fungal microsymbionts are beneficial soil organisms with an immense potential to resolve soil health problems. Among the known species, only a few have been characterized and identified. Their contribution includes rhizosphere health, pest and disease prevention, increased resource sharing and utilization, improved root activity efficiency, nutrient and water absorption, and oxygen and carbon dioxide balance in the root environment.

The importance of AMF in nutrient acquisition, cycling, water relations, bioprotection against pathogens, phytohormone production, and aggregate formation has been reported by Powlton et al. (2001) and Jeffries et al. (2003). Ryan and Graham (2002) reviewed the role of AMF in crops and concluded that AMF cannot replace mineral fertilizers but there was an increase in colonization whenever soils were poor in nutrients, especially P. Where poorly soluble phosphate rock was applied, for instance, AMF colonization significantly increased, suggesting a greater role for AMF in low–input agriculture systems with high degree of internal nutrient cycling than intensive production systems. AMF plays a role in soil health and fertility through complex microbial interactions in the rhizosphere. The success of AMF may imply changes in crop varieties, soil fertility inputs, and hence the production system and the output. There is paucity of information on how AFM technology can be effectively applied to enhance nutrient cycling in rice systems in which symbiosis occurs in approximately 80% of terrestrial plants.

5.4 ECONOMICS OF LEGUMES FOR EXTREMES OF WEATHER AND CLIMATE

Legumes have demonstrated immense contribution to nitrogen budgets at farm level, nutrition for both children and adults, and financial benefits through marketing of the grains and the resultant products. Their overall financial contribution is limited by the (1) limited area under cultivation (legumes are planted to less than 5% of the total cultivated land), (2) low-input investments on production, and (3) limited mechanization in production, harvesting, and postharvest processes. Furthermore, most small-holder farmers allocate potions with relatively degraded soil for legumes and the seed quality in legume value chain remains the weakest link.

With increased incidence of extremes of weather especially droughts, legumes are more threatened than ever before. Soybeans, for example, have been found to be more sensitive to drought and heat stresses than cowpeas and groundnuts. However, in the recent years, research on drought-tolerant cultivars has increased. More work is required on soybeans. Cowpeas and groundnuts seem to have cultivars that can tolerate drought and heat. An increased role for legumes is envisaged when suitable germplasm is coupled with good agronomic practices.

5.5 GAPS IN FUTURE RESEARCH

Several knowledge gaps still exist in the rhizobiology research for each of the microsymbiont. Across the three groups, i.e., legume–rhizobia symbiosis, mycorrhizal associations, and nonaflatoxigenic associations, more work is required to explore opportunities of crop cultivars that can become a single host for multiple organisms at the same time without causing conflicting responses. In most cases, data on population and species characterization for major soil groups in Southern Africa is scarce, a sign that not much is being done in this area. Simplified methods to determine the presence and/or absence of AMF, i.e., a quick assessment toolbox, will be handy for most practitioners. Spore germination conditions for improved infections need to be defined for major climatic zones. Specific issues for the individual groups are described in the following.

Legume-rhizobia symbiosis:

1. The legume genotype characteristics seem to have a dominant influence on the partnership with rhizobia. There is need to study new crop types and the associated management practices in-line, with the objective of increasing the N_2 fixation efficiency and the amount of N_2 fixed per unit biomass of legume crops under changing climate.

2. Rhizobia inoculant quality is a major determinant of effective symbiosis. Simplified methods of testing effectiveness at field level may result in detection of ineffective strains. Although the method of packaging rhizobia is well understood, this has been not translated into reduced incidence of distribution of materials with low rates of rhizobial population.

The relationship between quality of rhizobia products and weather parameters require attention.

3. Increased grain harvest index for the promiscuous legumes is an important area that required development. Majority of promiscuous cultivars tend to be highly vegetative and result in lower grain yield, a characteristic many farmers do not prefer. A common variety in Southern Africa originally bred in Zambia and called *Magoye* has wonderful vegetative growth habits but yields less when compared to specific varieties.

4. The P requirement for legume production has been highlighted in several research studies; however, the rates of application have not been determined for a whole range of legumes. There is need for renewed efforts in developing legume–specific fertilization strategies for major legumes, which can then be tailored to cultivars suitable for climate affected locations.

5. Other nutrient requirements for legumes need to be studied i.e., use of K and N together with micro-nutrients.

Mycorrhizal associations:

1. Most of the successfully established mycorrhizal associations seem to respond to a stimulus, e.g., low P or moisture content in soil, and this has limited the strategic planning of use of mycorrhiza because in the absence of the stimulating environment the infection area low. More work needs to be done to enable farmers to choose where fungal infections will enhance growth of the association.

2. Inoculating soils with low fungal population will assist in boosting the population and hence the probability of infection, leading to crop enhancement. The method of culturing and bulking and identifying suitable storage media is urgently required.

Nonaflatoxigenic associations:

1. The use of atoxigenic strains to manage aflatoxin levels in maize and groundnuts in Africa, and other parts of the world, will require an investment to optimize, adapt, and deploy the technology in a sustainable manner across different environments and users.

2. Given the large number of exploratory investigations in Africa, studies are needed to evaluate the impact of the low rate of genetic recombination, which will then enable the deployment of the technology in diverse settings.

3. Soil and climatic factors which support mitigation of aflatoxins need to be analyzed across southern Africa.

5.6 CONCLUSIONS

The utilization of microsymbionts in driving sustainable intensification options in cropping systems needs closer attention to work out site specificities that are currently hampering wide applications. Improved management practices on legumes that focus on the efficient cultivar-inoculant-P combinations are required. Similarly inoculant for AMF and increased application of this technology has the potential to resolve water and nutrient stresses in drought-prone areas affected by climate change and variability.

Since the outbreak of the turkey X disease, tremendous progress has been made in the control of aflatoxin contamination in crops. Use of nonaflatoxigenic strains of *A. flavus* to competitively exclude aflatoxin-producing strains has emerged as the best management practice for reducing aflatoxin concentrations and has led to the development of commercially registered products, first in cotton (2003) and then in groundnut (2004) and maize (2008). Multiagency, multistate, large-scale field trials are underway to determine the extent of aflatoxin level reduction that is possible and how to best incorporate biocontrol into standard agricultural practices. Major efforts are required to develop and apply the biocontrol technology and bring it to smallholder farmers.

REFERENCES

Abaidoo, R.C., Keyser, H.H., Singleton, P.W., Dashiell, K.E., Sanginga, N., 2007. Population size, distribution, and symbiotic characteristics of indigenous *Bradyrhizobium* spp. that nodulate TGx soybean genotypes in Africa. Applied Soil Ecology 35, 57–67. http://dx.doi.org/10.1016/j.apsoil.2006.05.006.

Abbas, H.K., Weaver, M.A., Horn, B.W., Carbone, I., Monacell, J.T., Shier, W.T., 2011. Selection of *Aspergillus flavus* isolates for biological control of aflatoxins in corn. Toxin Reviews 30 (2–3), 59–70.

Abbas, H.K., Weaver, M.A., Zablotowicz, R.M., Horn, B.W., Shier, W.T., 2005. Relationships between aflatoxin production and sclerotia formation among isolates of Aspergillus section Flavi from the Mississippi Delta. European Journal of Plant Pathology 112 (3), 283–287.

Abbas, H.K., Wilkinson, J.R., Zablotowicz, R.M., Accinelli, C., Abel, C.A., Bruns, H.A., Weaver, M.A., 2009. Ecology of Aspergillus flavus, regulation of aflatoxin production, and management strategies to reduce aflatoxin contamination of corn. Toxin Reviews 28 (2–3), 142–153.

Abbas, H.K., Zablotowicz, R.M., Bruns, H.A., Abel, C.A., 2006. Biocontrol of aflatoxin in corn by inoculation with non-aflatoxigenic Aspergillus flavus isolates. Biocontrol Science and Technology 16 (5), 437–449.

Abate, T., Alene, A.D., Bergvinson, D., Shiferaw, B., Silim, S., Orr, A., Asfaw, S., 2012. Tropical Grain Legumes in Africa and South Asia: Knowledge and Opportunities. International Crops Research Institute for the Semi-arid Tropics, Nairobi.

Adetuniji, M.C., Atanda, O.O., Ezekiel, C.N., Dipeolu, A.O., Uzochukwu, S.V.A., Oyedepo, J., Chilaka, C.A., 2014. Distribution of mycotoxins and risk assessment of maize consumers in five agro-ecological zones of Nigeria. European Food Research and Technology 239 (2), 287–296.

Agag, B.I., 2004. Mycotoxins in Foods and Feeds 1-Aflatoxins., 7. Assiut University Bulletin for Environmental Researches, pp. 173–205. http://www.aun.edu.eg/env_enc/env%20 mar/173-206.PDF.

Atehnkeng, J., Ojiambo, P.S., Cotty, P.J., Bandyopadhyay, R., 2014. Field efficacy of a mixture of atoxigenic Aspergillus flavus Link: Fr vegetative compatibility groups in preventing aflatoxin contamination in maize (Zeamays L.). Biological Control 72, 62–70.

Azziz-Baumgartner, E., Lindblade, K., Gieseker, K., Schurz-Rogers, H., Kieszak, S., Njapau, H., Schleicher, R., McCoy, L., Misore, A., DeCock, K., Rubin, C., Slutsker, L., the Aflatoxin Investigative Group, 2005. Case–control study of an acute aflatoxicosis outbreak, Kenya, 2004. Environmental Health Perspectives 113 (12), 1779–1783.

Bala, A., Karanja, N., Murwira, M., Lwimbi, L., Abaidoo, R., Giller, K., 2011. Production and Use of Rhizobial Inoculants in Africa (WWW Document) www.N2Africa.org.

Bhat, R.V., Krishnamachari, K.A., 1977. Follow-up study of aflatoxic hepatitis in parts of western India. Indian Journal of Medical Research 66, 55–58.

Bloem, J.F., Trytsman, G., Smith, H.J., 2009. Biological nitrogen fixation in resource-poor agriculture in South Africa. Symbiosis 48, 18–24. http://dx.doi.org/10.1007/ BF03179981.

Blount, W.P., 1961. Turkey "X" disease. Turkeys 9 (2), 52–55.

Bock, C.H., Cotty, P.J., 1999. Wheat seed colonized with atoxigenic Aspergillus flavus: characterization and production of a biopesticide for aflatoxin control. Biocontrol Science and Technology 9 (4), 529–543.

Boonen, J., Malysheva, S.V., Taevernier, L., Di Mavungu, J.D., De Saeger, S., De Spiegeleer, B., 2012. Human skin penetration of selected model mycotoxins. Toxicology 301 (1), 21–32.

Brown, R.L., Cotty, P.J., Cleveland, T.E., 1991. Reduction in aflatoxin content of maize by atoxigenic strains of Aspergillus flavus. Journal of Food Protection 54 (8), 623–626.

Cardwell, K.F., Henry, S.H., 2004. Risk of exposure to and mitigation of effect of aflatoxin on human health: a West African example. Journal of Toxicology: Toxin Reviews 23 (2–3), 217–247.

CAST (Council for Agriculture Science and Technology), 2003. Mycotoxins Risks in Plant, Animal, and Human Systems. Task Force Report, Ames, Iowa. 139.

Chang, P.-K., Horn, B.W., Dorner, J.W., 2005. Sequence breakpoints in the aflatoxin biosynthesis gene cluster and flanking regions in nonaflatoxigenic Aspergillus flavus isolates. Fungal Genetics and Biology 42, 914–923.

Chikowo, R., Mapfumo, P., Nyamugafata, P., Giller, K.E., 2004. Woody legume fallow productivity, biological N2-fixation and residual benefits to two successive maize crops in Zimbabwe. Plant and Soil 262 (1-2), 303–315.

Cole, R.J., Sanders, T.H., Hill, R.A., Blankenship, P.D., 1985. Mean geocarposphere temperatures that induce preharvest aflatoxin contamination of peanuts under drought stress. Mycopathologia 91 (1), 41–46.

Cotty, P.J., 1989. Virulence and cultural characteristics of two Aspergillus flavus strains pathogenic on cotton. Phytopathology 79, 808–814.

Cotty, P.J., 1994. Comparison of four media for the isolation of Aspergillus flavus group fungi. Mycopathologia 125, 157–162.

Cotty, P.J., 1997. Aflatoxin-producing potential of communities of Aspergillus section Flavi from cotton producing areas in the United States. Mycological Research 101, 698–704.

Cotty, P.J., 2006. Biocompetitive exclusion of toxigenic fungi. The mycotoxin factbook: food and feed topics 179–197.

Cotty, P.J., Antilla, L., Wakelyn, P.J., 2007. Competitive exclusion of aflatoxin producers: farmer-driven research and development. In: Vincent, C., Goettel, M.S., Lazarovits, G. (Eds.), Biological Control: A Global Perspective. CAB International, Oxfordshire, UK, pp. 241–253.

Cramer, M.D., Van Cauter, A., Bond, W.J., 2010. Growth of N2-fixing African savanna Acacia species is constrained by below-ground competition with grass. Journal of Ecology 98, 156–167. http://dx.doi.org/10.1111/j.1365-2745.2009.01594.x.

Cuero, R., Ouellet, T., Yu, J., Mogongwa, N., 2003. Metal ion enhancement of fungal growth, gene expression and aflatoxin synthesis in Aspergillus flavus: RT-PCR characterization. Journal of Applied Microbiology 94 (6), 953–961.

Das, I.K., Fakrudin, B., Arora, D.K., 2008. RAPD cluster analysis and chlorate sensitivity of some Indian isolates of Macrophomina phaseolina from sorghum and their relationships with pathogenicity. Microbiological Research 163 (2), 215–224.

Dorner, J.W., 2004. Biological control of aflatoxin contamination of crops. Journal of Toxicology: Toxin Reviews 23 (2-3), 425–450.

Dorner, J.W., 2009. Biological control of aflatoxin contamination in corn using a nontoxigenic strain of Aspergillus flavus. Journal of Food Protection 72 (4), 801–804.

Dorner, J.W., Lamb, M.C., 2006. Development and commercial use of afla-guard, an aflatoxin biocontrol agent. Mycotoxin Research 22 (1), 33–38.

Feng, G.H., Leonard, T.J., 1998. Culture conditions control expression of the genes for aflatoxin and sterigmatocystin biosynthesis in Aspergillus parasiticus and A. nidulans. Applied and Environmental Microbiology 64 (6), 2275–2277.

Fening, J.O., Danso, S.K.A., 2002. Variation in symbiotic effectiveness of cowpea bradyrhizobia indigenous to Ghanaian soils. Applied Soil Ecology 21, 23–29. http://dx.doi.org/10.1016/S0929-1393(02)00042-2.

Forgacs, J., Carll, W.T., 1962. Mycotoxicoses. Adv. Vet. Science 7, 273–882.

Gachomo, E.W., Kotchoni, S.O., 2008. The use of Trichoderma harzianum and T. viride as potential biocontrol agents against peanut microflora and their effectiveness in reducing aflatoxin contamination of infected kernels. Biotechnology 7, 439–447.

Gan, L.S., Skipper, P.L., Peng, X., Groopman, J.D., Chen, J.S., Wogan, G.N., Tannenbaum, S.R., 1988. Serum albumin adducts in the molecular epidemiology of aflatoxin carcinogenesis: correlation with aflatoxin B1 intake and urinary excretion of aflatoxin M1. Carcinogenesis 9, 1323–1325.

Giller, K.E., 2001. Nitrogen Fixation in Tropical Cropping Systems, second ed. CABI Publishing, Wallingford.

Giller, K.E., Murwira, M.S., Dhliwayo, D.K.C., Mafongoya, P.L., Mpepereki, S., 2000. Soyabeans and sustainable agriculture Promiscuous soyabeans in Southern Africa. Fields Crop Research 65, 137–149. http://dx.doi.org/10.1016/S0378-4290(99)00083-0.

Gong, Y.Y., Hounsa, A., Egal, S., Turner, P.C., Sutcliffe, A.E., Hall, A.J., Cardwell, K., Wild, C.P., 2004. Postweaning exposure to aflatoxin results in impaired child growth: a longitudinal study in Benin, West Africa. Environmental Health Perspectives 112, 1334–1338.

Gong, Y.Y., Turner, P.C., Hall, A.J., Wild, C.P., 2008. Aflatoxin exposure and impaired child growth in West Africa: an unexplored international public health burden? In: Leslie, J.F., Bandyopadhyay, R., Visconti, A. (Eds.), Mycotoxins—Detection Methods, Management, Public Health and Agricultural Trade. CAB International, Oxfordshire, UK, pp. 53–66.

Groopman, J.D., Donahue, P.R., Zhu, J.Q., Chen, J.S., Wogan, G.N., 1985. Aflatoxin metabolism in humans: detection of metabolites and nucleic acid adducts in urine by affinity chromatography. Proceedings of the National Academy of Sciences 82 (19), 6492–6496.

Groopman, J.D., Hall, A.J., Whittle, H., Hudson, G.J., Wogan, G.N., Montesano, R., Wild, C.P., 1992a. Molecular dosimetry of aflatoxin-N 7-guanine in human urine obtained in the Gambia, West Africa. Cancer Epidemiology Biomarkers and Prevention 1, 221–227.

Groopman, J.D., Hasler, J.A., Trudel, L.J., Pikul, A., Donahue, P.R., Wogan, G.N., 1992b. Molecular dosimetry in rat urine of aflatoxin-N7-guanine and other aflatoxin metabolites by multiple monoclonal antibody affinity chromatography and immunoaffinity/high performance liquid chromatography. Cancer Research 52, 267–274.

Horn, B.W., 2003. Ecology and population biology of aflatoxigenic fungi in soil. Journal of Toxicology: Toxin Reviews 22 (2–3), 351–379.

Indieka, S., Odee, D., 2005. Nodulation and growth response of Sesbania sesban (L.) Merr. to increasing nitrogen(ammonium) supply under glasshouse conditions. African Journal of Biotechnology 4, 57–60.

IPCC, 2007. Climate change: impacts, adaptation and vulnerability. In: Parry, M.L., Canziani, O.F., Palutikof, J.P., van der Linden, P.J., Hanson, C.E. (Eds.), Contribution of Working Group II to the Fourth Assessment Report of the Intergovernmental Panel on Climate Change. Cambridge University Press, Cambridge.

Jeffries, P., Gianinazzi, S., Perotto, S., Turnau, K., Barea, J.M., 2003. The contribution of arbuscular mycorrhizal fungi in sustainable maintenance of plant health and soil fertility. Biology and fertility of soils 37 (1), 1–16.

Jiang, Q., Wang, Y., Hao, Y., Juan, L., Teng, M., Zhang, X., Li, M., Wang, G., Liu, Y., 2009. miR2Disease: a manually curated database for microRNA deregulation in human disease. Nucleic acids research 37 (Suppl. 1), D98–D104.

Krishnamachari, K.A., Bhat, R.V., Nagarajan, V., Tilak, T.B., 1975. Hepatitis due to aflatoxicosis. An outbreak in Western India. Lancet 1, 1061–1063.

Lemaire, B., Dlodlo, O., Chimphango, S., Stirton, C., Schrire, B., Boatwright, J.S., Honnay, O., Smets, E., Sprent, J., James, E.K., Muasya, A.M., 2015. Symbiotic diversity, specificity and distribution of rhizobia in native legumes of the Core Cape Subregion (South Africa). FEMS Microbiology Ecology 91, 1–42. http://dx.doi.org/10.1093/femsec/fiu024.

Lewis, L., Onsongo, M., Njapau, H., Schurz Rogers, H., Luber, G., Kieszak, S., Nyamongo, J., Backer, L., Dahiye, A., Misore, A., DeCock, K., Rubin, C., Year and the Kenya Aflatoxicosis Investigation Group, 2005. Aflatoxin contamination of commercial maize products during an outbreak of acute aflatoxicosis in eastern and Central Kenya. Environmental Health Perspectives 113 (12), 1763–1767.

Lyn, M.E., Abbas, H.K., Zablotowicz, R.M., Johnson, B.J., 2009. Delivery systems for biological control agents to manage aflatoxin contamination of pre-harvest maize. Food Additives and Contaminants 26 (3), 381–387.

Mafongoya, P.L., Bationo, A., Kihara, J., Waswa, B.S., 2006. Appropriate technologies to replenish soil fertility in Southern Africa. Nutrient Cycling in Agroecosystems 76, 137–151. http://dx.doi.org/10.1007/s10705-006-9049-3.

Maingi, J.M., Gitonga, N.M., Shisanya, C.A., Hornetz, B., Muluvi, G.M., 2006. Population levels of indigenous bradyrhizobia nodulating promiscuous soybean in two Kenyan soils of the semi-arid and semi-humid agroecological zones. Journal of Agriculture and Rural Development in the Tropics and Subtropics 107, 149–159.

Mapfumo, P., Mpepereki, S., Mafongoya, P., 2000. Pigeonpea rhizobia prevalence and crop response to inoculation in Zimbabwean smallholder-managed soils. Experimental Agriculture 36 (04), 423–434.

Mc Millian, W.W., Wilson, D.M., Widstrom, N.W., 1985. Aflatoxin contamination of preharvest corn in Georgia: a six-year study of insect damage and visible Aspergillus flavus. Journal of Environmental Quality 14 (2), 200–202.

Mehl, H.L., Jaime, R., Callicott, K.A., Probst, C., Garber, N.P., Ortega-Beltran, A., Grubisha, L.C., Cotty, P.J., 2012. Aspergillus flavus diversity on crops and in the environment can be exploited to reduce aflatoxin exposure and improve health. Annals of the New York Academy of Sciences 1273 (1), 27–17.

Mohammadipour, M., Mousivand, M., Salehi Jouzani, G., Abbasalizadeh, S., 2009. Molecular and biochemical characterization of Iranian surfactin-producing Bacillus subtilis isolates and evaluation of their biocontrol potential against Aspergillus flavus and Colletotrichum gloeosporioides. Canadian Journal of Microbiology 55 (4), 395–404.

Mulas, D., Seco,V., Casquero, P.A.,Velázquez, E., González-Andrés, F., 2015. Inoculation with indigenous rhizobium strains increases yields of common bean (*Phaseolus vulgaris* L.) in Northern Spain, although its efficiency is affected by the tillage system. Symbiosis 67, 113–124. http://dx.doi.org/10.1007/s13199-015-0359-6.

Niessen, L., 2007. PCR-based diagnosis and quantification of mycotoxin producing fungi. International Journal of Food Microbiology 119 (1), 38–46.

Ngindu, A., Johnson, B.K., Kenya, P.R., Ngira, J.A., Ocheng, D.M., Nandwa, H., Omondi, T.N., Jansen, A.J., Ngare, W., Kaviti, J.N., et al., 1982. Outbreak of acute hepatitis caused by aflatoxin poisoning in Kenya. Lancet 1, 1346–1348.

Olarte, R.A., Horn, B.W., Dorner, J.W., Monacell, J.T., Singh, R., Stone, E.A., Carbone, I., 2012. Effect of sexual recombination on population diversity in aflatoxin production by Aspergillus flavus and evidence for cryptic heterokaryosis. Molecular Ecology 21 (6), 1453–1476.

Omondi, J.O., Mungai, N.W., Ouma, J.P., Patrick, F., 2014. Effect of tillage on biological nitrogen fixation and yield of soybean (*Glycine max* L. Merril) varieties. Australian Journal of Crop Science 8, 1140–1146.

Payne, G.A., 1998. Process of contamination by aflatoxin-producing fungi and their impact on crops. Mycotoxins in Agriculture and Food Safety 9, 279–306.

Payne, G.A., Brown, M.P., 1998. Genetics and physiology of aflatoxin biosynthesis. Annual Review of Phytopathology 36 (1), 329–362.

Poirier, M.C., Santella, R.M., Weston, A., 2000. Carcinogen macromolecular adducts and their measurement. Carcinogenesis 21, 353–359.

Powlton, D.S., Hirsch, P.R., Brookes, P.C., 2001. The role of soil microorganisms in soil organic matter conservation in the tropics. Nutrient Cycling in Agroecosystems 61, 41–51.

Probst, C., Njapau, H., Cotty, P.J., 2007. Outbreak of an acute aflatoxicosis in Kenya in 2004: identification of the causal agent. Applied and Environmental Microbiology 73, 2762–2764.

Rachel, B.K., Snapp, S., Chirwa, M., Shumba, L., Msachi, R., 2007. Participatory research on legume diversification with Malawanian smallholder farmers for improved human nutrition and soil fertility. Experimental Agriculture 43, 437–453. http://dx.doi.org/10.1017/S0014479707005339.

Richard, J.L., 2007. Some major mycotoxins and their mycotoxicoses – an overview. International Journal of Food Microbiology 119, 3–10.

Rowe, E., Giller, K.E., 2003. Legumes for soil fertility in southern Africa: Needs, potential and realities. In: Grain Legumes and Green Manures for Soil Fertility in Southern Africa: Taking Stock of Progress. Soil Fert Net-CIMMYT, Harare, Zimbabwe, pp. 15–19.

Ryan, M.H., Graham, J.H., 2002. Is there a role for arbuscular mycorrhizal fungi in production agriculture? In: Diversity and Integration in Mycorrhizas. Springer, Netherlands, pp. 263–271.

Santella, R.M., 1999. Immunological methods for detection of carcinogen-DNA damage in humans. Cancer Epidemiology Biomarkers & Prevention 8 (9), 733–739.

Sobowale, A.O., Olurin, T.O., Oyewole, O.B., 2007. Effect of lactic acid bacteria starter culture fermentation of cassava on chemical and sensory characteristics of fufu flour. African Journal of Biotechnology 6 (16).

Strosnider, H., Azziz-Baumgartner, E., Banziger, M., Bhat, R.V., Breiman, R., Brune, M.N., DeCock, K., Dilley, A., Groopman, J., Hell, K., et al., 2006. Workgroup report: public health strategies for reducing aflatoxin exposure in developing countries. Environmental Health Perspectives 114, 1898–1903.

Sweany, R.R., Damann Jr., K.E., Kaller, M.D., 2011. Comparison of soil and corn kernel *Aspergillus flavus* populations: evidence for niche specialization. Phytopathology 101 (8), 952–959.

Turner, P.C., Moore, S.E., Hall, A.J., Prentice, A.M., Wild, C.P., 2003. Modification of immune function through exposure to dietary aflatoxin in Gambian children. Environmental Health Perspectives 111 (2), 217.

Van Kessel, C., Hartley, C., 2000. Agricultural management of grain legumes: has it led to an increase in nitrogen fixation? Field Crops Research 65, 165–181. http://dx.doi.org/10.1016/S0378-4290(99)00085-4.

Wicklow, D.T., Roth, S., Deyrup, S.T., Gloer, J.B., 2005. A protective endophyte of maize: *Acremonium zeae* antibiotics inhibitory to *Aspergillus flavus* and *Fusarium verticillioides*. Mycological Research 109 (5), 610–618.

Widstrom, N.W., McMillian, W.W., Beaver, R.W., Wilson, D.M., 1990. Weather-associated changes in aflatoxin contamination of preharvest maize. Journal of Production Agriculture 3 (2), 196–199.

Wild, C.P., Turner, P.C., 2002. The toxicology of aflatoxins as a basis for public health decisions. Mutagenesis 17, 471–481.

Wild, C.P., Hudson, G.J., Sabbioni, G., Chapot, B., Hall, A.J., Wogan, G.N., Whittle, H., Montesano, R., Groopman, J.D., 1992. Dietary intake of aflatoxins and the level of albumin-bound aflatoxin in peripheral blood in the Gambia, West Africa. Cancer Epidemiology Biomarkers and Prevention 1, 229–234.

Williams, J., Phillips, T.D., Jolly, P.E., Stiles, J.K., Jolly, C.M., Aggarwal, D., 2004. Human aflatoxicosis in developing countries: a review of toxicology, exposure, potential health consequences, and interventions. American Journal of Clinical Nutrition 80, 1106–1122.

Wu, F., Khlangwiset, P., 2010. Health economic impacts and cost-effectiveness of aflatoxin-reduction strategies in Africa: case studies in biocontrol and post-harvest interventions. Food Addit Contam Part A Chem Anal Control Expo Risk Assess 27 (4), 496–509.

Wu, F., Narrod, C., Tiongco, M., Liu, L., Collier, W., Scott, R., 2012. The Economic Impact of Health Consequences of Aflatoxin: Estimation of the Global Burden of Disease (Aflacontrol Working Paper, IFPRI).

FURTHER READING

Abbas, H.K., Zablotowicz, R.M., Horn, B.W., Phillips, N.A., Johnson, B.J., Jin, X., Abel, C.A., 2012. Comparison of major biocontrol strains of non-aflatoxigenic *Aspergillus flavus* for the reduction of aflatoxins and cyclopiazonic acid in maize. Food Additives and Contaminants Part A 28, 198–208.

Accinelli, C., Saccà, M.L., Abbas, H.K., Zablotowicz, R.M., Wilkinson, J.R., 2009. Use of a granular bioplastic formulation for carrying conidia of a non-aflatoxigenic strain of *Aspergillus flavus*. Bioresource Technology 100, 3997–4004.

Bandyopadhyay, R., Kiewnick, S., Atehnkeng, J., Donner, M., Cotty, P.J., Hell, K., 2005. Biological control of aflatoxin contamination in maize in Africa. In: Presented in the Conference on International Agricultural Research for Development, Tropentag.

Brown, R.L., Chen, Z.Y., Menkir, A., Cleveland, T.E., Cardwell, K., Kling, J., et al., 2001. Resistance to aflatoxin accumulation in kernels of maize inbreds selected for ear rot resistance in West and Central Africa. Journal of Food Protection 64 (3), 396–400.

Chang, P.K., Hua, S.S.T., 2007. Nonaflatoxigenic *Aspergillus flavus* TX9-8 competitively prevents aflatoxin accumulation by *A. flavus* isolates of large and small sclerotial morphotypes. International Journal of Food Microbiology 114 (3), 275–279.

Chang, P.K., Bhatnagar, D., Cleveland, T.E., Bennett, J.W., 1995. Sequence variability in homologs of the aflatoxin pathway gene aflR distinguishes species in *Aspergillus* section Flavi. Applied and Environmental Microbiology 61, 40–43.

Cotty, P.J., Bandyopadhyay, R., 2010. Working together: partnering with grower organizations from development through distribution to make aflatoxin biocontrol a reality in the US/AfricaAspergillus. Phytopathology 100, S162.

Cotty, P.J., Cardwell, K.F., 1999. Divergence of West African and North American Communities of *Aspergillus* section Flavi. Applied and Environmental Microbiology 65, 2264–2266.

Cotty, P.J., Bayman, P., Egel, D.S., Elias, K.S., 1994. Agriculture, aflatoxins, and *Aspergillus*. In: Powell, K.A., Renwick, A., Perberdy, J. (Eds.), The Genus *Aspergillus*: From Taxonomy and Genetics to Industrial Application. Plenum Press, New York, pp. 1–27.

Cotty, P.J., Probst, C., Jaime-Garcia, R., 2008. Etiology and management of aflatoxin contamination. In: Leslie, J.F., Bandyopadhyay, R., Visconti, A. (Eds.), Mycotoxins: Detection Methods, Management Public Health and Agricultural Trade. CAB International, Oxfordshire, UK, pp. 287–299.

Daniel, J.H., Lewis, W.L., Redwood, A.Y., Kieszal, S., Breiman, F.R., Flanders, D., Bell, C., Mwihia, J., Ogana, G., Likimani, S., Straetemans, M., McGeehin, A.M., 2011. Comprehensive assessment of maize aflatoxin levels in eastern Kenya, 2005-2007. Environmental Health Perspectives 119, 1794–1799.

De Groote, H., Bett, C., Okuro, J., Odedo, M., Mose, L., Wekesa, E., 2002. Direct estimation of maize crop losses due to stem borers in Kenya, preliminary result from 2000 and 2001. In: Friesen, D.K., Palmer, A.F.E. (Eds.), Integrated Approaches to Higher Maize Productivity in the New Millennium. Proceedings of the 7th Eastern and Southern Africa Regional Maize Conference, Nairobi, Kenya, February 11–15, 2002. CIMMYT, Mexico, DF, pp. 401–406.

Donner, M., Atehnkeng, J., Sikora, R.A., Bandyopadhyay, R., Cotty, P.J., 2009. Distribution of *Aspergillus* section Flavi in soils of maize fields in three agroecological zones of Nigeria. Soil Biology and Biochemistry 41, 37–44.

Forgacs, J., 1962. Mycotoxicoses—the neglected diseases. Feedstuffs 34, 124–134.

Kensler, T.W., Roebuck, B.D., Wogan, G.N., Groopman, J.D., 2011. Aflatoxin: a 50-Year Odyssey of Mechanistic and Translational toxicology. Toxicological Sciences 120 (S1), S28–S48.

Koide, R.T., Mosse, B., 2004. A history of research on arbuscular mycorrhiza. Mycorrhiza 14, 145–163.

Marechera, G., Ndwiga, J., 2015. Estimation of the potential adoption of Aflasafe among smallholder maize farmers in lower Eastern Kenya. African Journal of Agricultural and Resource Economics 10 (1), 72–85.

Mehl, H.L., Cotty, P.J., 2010. Variation in competitive ability among isolates of *Aspergillus flavus* from different vegetative compatibility groups during maize infection. Phytopathology 100, 150–159.

Mehl, H.L., Cotty, P.J., 2011. Influence of the host contact sequence on the outcome of competition among *Aspergillus*. Applied and Environmental Microbiology.

Moss, M.O., 2002. Risk assessment for aflatoxins in foodstuffs. International Biodeterioration and Biodegradation 50 (3–4), 137–142.

Robens, J., Cardwell, K.F., 2005. The costs of mycotoxin management in the United States. In: Abbas, H.K. (Ed.), Aflatoxin and Food Safety (Food Science and Technology) Available at: http://www.cabdirect.org/abstracts/20063046008.html.

Robens, J., Cardwell, K.F., 2003. The costs of mycotoxin management to the USA: management of aflatoxins in the United States. Journal of Toxicology: Toxin Reviews 22, 139–152.

CHAPTER 6

Reducing Risk of Weed Infestation and Labor Burden of Weed Management in Cropping Systems

Mohamed J. Kayeke[1], Nhamo Nhamo[2], David Chikoye[2]

[1]Mikocheni Agricultural Research Institute (MARI), Dar es Salaam, Tanzania; [2]International Institute of Tropical Agriculture (IITA), Southern Africa Research and Administration Hub (SARAH) Campus, Lusaka, Zambia

Contents

Smart Technologies for Sustainable Smallholder Agriculture
ISBN 978-0-12-810521-4
http://dx.doi.org/10.1016/B978-0-12-810521-4.00006-2

6.1 INTRODUCTION

Weeds are the most underestimated agricultural pests and yet infestations are responsible for increased cost of crop production. Climate change is bound to increase weed pressure in cropping systems as several weed species expand their range under warmer conditions and variable rainfall patterns. Peters et al. (2014) have described the effects of climate change on weeds and linked them to three shifts, which will occur across various scales, i.e., first range shifts at the landscape scale, second niche shifts at the community scale, and the third trait shifts of individual species at the population scale. Therefore, the spread of weed species in arable habitats will be influenced by the climatic variability and anthropogenic factors. More specifically species-specific trait response to climate variability and change will define the impact of weeds on cropping systems (Clements et al., 2014). The overall cost of weeds will change in relation to the ecological, biological, and evolutionary response to climate change and the management practices employed by the majority of farmers.

Weeds compete with crop for nutrients, water, space, and light. They are spontaneous, can be harmful to crop, humans, or animals; therefore, they interfere directly with crop value chains and production operations. For instance, weed species such as *Parthenium hysterophorus* has strong allelopathic thus impacts crops, human, and animals. Majority of weed species have adaptation mechanisms that allow them to survive a wide range of climatic, edaphic, and biotic constrains and as a result they are difficult to control. The resultant risks of having a weedy cropping field is a loss of harvest, poor quality crop products, and interference with the process of harvesting. This translates into increased cost of production and negative environmental effects.

Labor- and energy-intensive manual methods, mechanical, cultural, and chemical weed management methods have their merits highlighted by positive research results in various cropping systems. Similarly, ecological methods of weed management are largely designed based on the ecological and biological traits of the weed species. Combinations of methods have been used to target the weed seed bank to control weed species composition in the long run. We define weed seed bank as the reserve of viable weed seeds present on the soil surface or in the soil (Chauhan and Gill, 2014). Depletion of weed seed bank is often through germination and then control, predation, or death through decay. Use of tillage, mulch, weed competitive cultivars, allelopathy, and early canopy development are important methods employed in weed management and control. The comparative advantages and effectiveness of these methods of weed control under climate change are not clear.

The world over weed management cost about $400 billion annually (Clements et al., 2014). On smallholder farms in southern Africa, weeding constitute an average of 30–60% of the labor input of all operations and the cost of labor invested in weeding cropping lands using mainly cultural weed management methods runs to millions per year, a financial burden borne by farm families. The cost of yield losses emanating from weeds have been estimated to be $1.45 billion in rice systems (Rodenburg and Johnson, 2009), about $3.41 billion in maize-based systems and $0.98 billion in cassava systems. In total, Africa stands to lose about $6 billion due to weed infestations in cropping systems and food production processes. Both climate variability and weed management practices account for the overall losses farmers incur during production and processing of agricultural products (Rosenzweig et al., 2001). In response to the huge cost and the difficulties associated with weed management there is a need for development of smart weed management practices that address the current and future challenges on smallholder farms. Development of these weed management practices may imply adapting some of the common farming systems in order to tackle the impacts of climate extremes and variability.

Weeds also affect the environmental quality, in particular, the water quality in water ways and storage facilities. Water hyacinth (*Eichhornia crassipes*) and *Salvinia* spp. are both notorious weeds associated with water bodies, which have taken chemical and biological control methods to contain the spread to other areas (Charudattan, 1986; Mitchell, 1972). Both the Kariba weed, *Salvinia molesta* (Mitchell), and the water hyacinth, *E. crassipes* (Mart.), have been reported in Kariba Dam and other inland manmade water bodies (Chikwenhere and Keswani, 1997; Masifwa et al., 2001). Even with the current control methods in place climate change may increase the range of these species and hence improved solutions will be required in the future.

Large water bodies, e.g., inland lakes, require large amounts of money to procure biological and chemical weed control inputs to keep the water clean and sustain the ecosystem.

Weeds outside crop production can be viewed as plants, which have numerous beneficial roles ranging from medicinal, soil fertility amendments, pest control capability, and aesthetic (decorative) values. For instance, *Ageratum conyzoides* has been highlighted for its medicinal and bioherbicidal applications (Xuan et al., 2004). Weeds can also control soil erosion (live mulching) and support ecological services assisting insects to complete their life cycles. In other cases, weeds grown systematically within a cropping system can be beneficial to the crop and environment, for example, *Mimosa invisa* and *Crotalaria ochroleuca* rotated with rainfed rice (Kayeke et al., 2007); *Desmodium* spp. and Napier grass in a push–pull technology (Cook et al., 2006; Khan et al., 2010; Midega et al., 2009). In the two situations, the particular weed species reduces infestation of other species through the crowding effects, suppression of germination, killing of weed seeds (suicidal germination), and improved soil fertility.

There are opportunities of using species-specific knowledge on evolution, biology, and ecology to develop early warning systems in view of climate variability. Given that weed species adapt to multiple pressures, e.g., heat and moisture, the development of bioclimatic models that can track the effect of multiyear droughts and heat waves on weed species distribution and assess the inversion risk will be a huge milestone toward understanding refining weed management practices (Clements et al., 2014). An early warning system and adaptive strategies for the dominant cropping systems in southern Africa will save millions of dollars.

This chapter highlights the progress made to date in weed management interventions suitable for smallholder farmers and the application of the technologies and explore opportunities of improving the use of weed management practices in the face of climate variability and change. We hypothesize that smart weed management techniques will reduce cost and labor demands, and improved environmental effects, therefore, reduce risk of weeds on agricultural production in the changing climate. The chapter discusses advances in weed management and is organized into five sections. Section 6.2 discusses the weed–crop interactions on smallholder farms; Section 6.3 looks at the effect of environmental factors influencing weed distribution; Section 6.4 describes suitable approaches in managing weeds in cropping systems; Section 6.6 looks at the research gaps; and Section 6.7 is the conclusion.

6.2 WEED-CROP INTERACTIONS ON SMALLHOLDER FARMS

Climate change will affect species dominance in different ecologies and this will determine which management practice to apply. Weeds are among the most significant biological yields reducing factors together with pests and disease and where infestations are severe yield losses as high as 90% have been recorded. Naturally weeds compete intensely with any plant and the competition is equally high among different weed species. In southern Africa there is a wide diversity of weed flora that colonizes agricultural land (both cropped land and livestock pastures). Under conditions where rainfall is limiting and temperature is unfavorable to crops. Some weed species dominate and utilize resources for vegetative and reproductive growth stages ahead of crops. Table 6.1 shows most common weed species found on arable land in southern Africa. Although absolute elimination of weeds is not necessary, the aim of the most effective management practice will be to reduce to bare minimum interference. There is a need for crop interference to be minimized, especially during the critical period of weed competition, a period where the effect of competition for resources should be least to avoid crop losses (Akobundu, 1987). This period varies with crop variety, cropping system, weed species, and climatic condition of a particular area. Climatic conditions that favor dominant weed flora development will lead to increased interference with crop growth.

Interference is defined here as the detrimental effects of a plant species on another as a result of their interaction, including both competition and allelopathy. In the same light competition is the relationship between two or more organisms in which the supply of a growth limiting factor falls below their combined demands. This is often the case for crops and weeds in soils with low soil fertility (due to mainly degraded

Table 6.1 Critical growth stages when weeding makes the most positive influence of crop production

Crop	Growth stage	Reproductive stage
Cassava	Early juvenile and harvesting	Bulking
Maize	Early juvenile	Silking, flowering, and grain filling
Rice	Early juvenile	Tillering, boating, grain filling
Sorghum	Early juvenile	Heading and grain filling

sandy soils) and semiarid areas (resulting from soil moisture deficiencies). Three major factors that influence competition are (1) weed factors (weed species, density, onset, and duration of crop–weed interaction), (2) crop factors (crop type, seeding rate, spatial arrangement, and architecture and availability growth factors), and (3) environmental factors [water and light (climate), tillage, soil fertility, and ground cover]. Competition detrimental to crop is enhanced when there is poor timing of application of weed management in the critical period of weed competition and at reduced frequency of weeding. Poor timing of operations coupled with irregular rainfall patterns disrupt weed management practices planned at the farm level. In rice systems, critical times for weeding have been identified generally as a period before tillering, booting, and flowering as it varies with planting system (direct seeding/transplanted) ecology (flooded/irrigated/rainfed). The situation is invariably the same for cassava, maize, and sorghum (Table 6.1).

6.2.1 Noxious Weeds

Weeds that parasitize crops form an important branch of weed science referred to as parasitic or noxious weeds. There are different groups of parasitic weeds including holoparasites (totally parasite), obligate parasites (growth and development depend on the host), and facultative parasite (semiparasite) (Parker and Riches, 1993). Research results have shown a considerable amount of progress in understanding their physiology and biochemistry behavior. Key weed species in this category are *Striga* spp., *Rhamphicarpa fistulosa*, *Alectra* spp., and *Orobanche* spp. (Rodenburg et al., 2015; Kayeke et al., 2010). A strong relationship exists between *Striga* spp. parasitism on maize, for instance, and low soil fertility status, in particular, nitrogen deficiency (Emechebe et al., 2004; Kamara et al., 2008). Addition of nitrogen has been used as a solution to reduce the effect of *Striga* spp. on maize.

Weeds from the Orobanchaceae family are serious parasitic pests in crop production, e.g., *Striga asiatica* (L.) Kuntze, *Striga hermonthica* (Del.) Benth and *S. aspera* (Wild) Benth (Rodenburg and Johnson, 2009). There is another group of weeds that produce chemicals that are detrimental to crops such as *P. hysterophorus* with its allelopathic effect but *Desmodium* spp. has allelopathic effect on other weeds, e.g., *S. asiatica* and *S. hermonthica*. Allelopathic effect of *Desmodium* spp. to *Striga* spp. has been utilized to develop a push–pull technology, which is a climate smart weed management technology (Khan et al.,

2011; Pickett et al., 2014). Under the push and pull technologies the weeds are used to interfere with germination of parasitic weeds to enhance crop growth and they provide competition useful for increasing growth.

6.3 ENVIRONMENTAL FACTORS INFLUENCING WEED DISTRIBUTION

Besides the weed factors and the crop factors, environmental factors have been found to correlate with growth and persistence of some weed species on the farms. The main environmental factors considered important in the study of weeds are climate, tillage methods, soil fertility status, and ground cover. Air quality has become an important factor in light of climate change and increased greenhouse gas (GHG) emissions. Among the GHGs, carbon dioxide (CO_2) stands out as the potential source of varied response from both crop and weeds on cropping lands. In a broad sense climate change effects on weeds will be driven by factors that can be grouped into the following: rainfall variability, increased temperature extremes, and increased CO_2 emissions.

6.3.1 Rainfall

Climate variability and change have created favorable environment for some weeds and the same conditions are unfavorable to others. In the course of this process there are chances of some weed species to spread to currently noninfested areas and thereby attack crops. This can be seen when there is reduced rainfall and extended dry spells in an environment favorable for *Striga* spp. infestation, in rainfed crops. Low humidity caused by dry weather conditions and high temperatures creates favorable environmental conditions for high transpiration from plants; weeds have a more advanced mechanism to parasitize other plants by accessing more moisture and nutrients from the soil (Parker and Riches, 1993). Perpetual conditions as a result of climate change will enable weed species such as *Striga* to spread into areas where it is not currently found. On the other hand, increased rainfall and flooding, a condition unfavorable for weeds such as *Oryza longistaminata* and *R. fistulosa* (a semiparasitic weed), which prefer hydromorphic conditions will alter prevalence. Rodenburg et al. (2015), Kayeke et al. (2010), and Kabiri et al. (2015) reported the transition between *Striga* and *Rhamphicarpa* ecologies, which keeps shifting yearly from dry to hydromorphic as a result of moisture level from one ecology to another. Therefore, the increased rainfall and moisture in the soil to a hydromorphic condition in an area extend the ecology for *R. fistulosa* while shrinking the ecology for *S. asiatica* in a rainfed system. Similar relationships between soil moisture and rainfall variability will affect both prevalence and species composition.

6.3.2 Temperature

Weeds like other plants are directly and indirectly affected by temperature in growth and development. The direct effects are on physiological plant functions such as photosynthesis, processes controlling translocation of nutrients, water, and photosynthates and respiration (Azam-Ali and Squire, 2002). Indirectly, temperature determines the overall growth rate, rate of biochemical reactions, and internal nutrient availability and activity (Mengel and Kirkby, 1987). Under increased temperature, the response of weeds to herbicide application changes. The control of weeds may be reduced due to limited effectiveness of the active ingredients in herbicide leading to herbicide resistance. Herbicide resistance can then be linked to climate change extreme events. Under increased temperatures, for instance, plant biochemical activities increase and interfere with the performance of the herbicide through decomposition or deactivation (Jablonkai, 2015).

6.3.3 Increased CO_2 Emission

Increased CO_2 in air can favor C4 and C3 weeds. According to Bunce (1993) an increase in CO_2 in the air reduces stomatal conductance and this can increase growth and development through the reduction of water loss from the plant; also reduced stomatal conductance can cause an increase in weed tolerance against herbicides in C3 and C4 plants (Ziska et al., 1999) through reduced uptake. With increase in greenhouse gas emissions, weed species that respond positively to high CO_2, for instance, will gain prominence compared to the species that are air quality neutral. The photosynthetic path taken by a plant distinguishes them between two groups, i.e., C3 and C4 plants. With increasing climate change affecting CO_2 concentration in the atmosphere due to increased emissions from fossil fuel burning, the weed floras seem to respond differently to herbicides application. The C4 group of weeds constitute the majority and the most difficult weeds to control in the world as they respond positively to increased temperature. The weed flora found on the majority of cropping lands is characterized by the presence of both grasses and broadleaves, annual and perennial weeds, upland and lowland species, species that are responsive and those that are neutral to CO_2 content in the atmosphere.

Therefore, climate change may lead to ecosystems disruptions, e.g., shifts or loss in dominant species, species redundancy, or loss in net ecosystems productivity. This suggests that the contribution of some species to productivity may be replaced by another set of species (Gitay et al., 2002). Only a few studies have looked at the combined effect of CO_2, drought, and temperature on both weeds and crops (Rodenburg and Johnson, 2009). Environmental factors, therefore, have the potential to influence the interaction between weeds and crops, determine the total biomass (weed and crops) per unit area and the long-term weeds ecology at catchment level.

6.3.4 Water and Light

Climate extremes in southern Africa have been associated with intermittent and terminal droughts and also extended mid-season droughts due to poor rainfall distribution. False start of a cropping season leads to huge financial losses as inputs are wasted and the length of the crop growing season is reduced. Unpredictable rainfall pattern impacts negatively on the weed management on smallholder farms that often combine manual, mechanical, and chemical methods. However, where water management through irrigation, timing of moist periods can be done to avoid the weed burden. Flooding can be used as a weed control method and only those species that can withstand flooded root zones will need additional attention (a common practice in rice systems). Whereas allowing for drying and increased aeration will increase the incidence of the upland type of weeds. Mechanical weeding can be timed to coincide with the drier soil profiles and hence making control much easier. Similarly, chemical weeding will be most effective on moist rather than dry conditions. Adequate moisture content in soils generally cause weed seed to germinate and a planned response to the flush of weeds following supply of moisture will be necessary.

Tropical weeds respond to light and light intensity with enhanced phenological development during the period when availability is high. Light stimulates early juvenile growth, germination, and flowering and seeding for some weed species. In mixed cropping systems weeds compete with crops for light and other resources. Little work has been done on photoperiodism on common weed species found on smallholder farms. Such work can increase accuracy of weed management practices if the photoperiod requirements for major weeds were used in the design of weed management. Light is related to heat and the effect of temperature on growth is well established.

6.3.5 Tillage and Implements

Conventional ploughing is a common feature in southern Africa communities where draught power from livestock is available. Although the principal objective of conventional ploughing would be to provide a good seedbed for the crops in the majority of cases it has also been used as a preliminary weed control method. Despite this notion, ploughing is nevertheless an effective method of spreading weed seeds across the fields and also burying, therefore preserving some in the subsoils. Seasonal application of similar tillage practices has a tendency to smoothen the dominant weed species for a given ecology making weed management easier to design. Delayed tillage and blind tillage have shown potential in reduction of weed infestation levels, hence avoidance of weeds infestation at early stages of crop growth (Akobundu, 1987). The two methods aim at destruction of first flush of weed before planting and destruction of the first flush before crop emergency, respectively. When the two methods are well practiced and associated with weed-tolerant varieties boosted with N fertilizers (Rodenburg and Johnson, 2013), the results are excellent.

6.3.6 Soil Fertility

Weeds, similar to domesticated crops, rely on soil nutrients to build biomass and where soils are fertile vigorous weed growth has been reported. The situation is different for *S. asiatica, S. hermonthica,* and *R. fistulosa* as these parasitic weeds are favored by low soil fertility especially low nitrogen levels in the soil. Therefore, improved soil fertility by organic or inorganic fertilizer will have a negative effect to such weeds, hence a component of integrated management. Application of N fertilizers has shown potential in reduction of infestation levels for parasitic weeds *Striga* spp. and *R. fistulosa* (Kayeke et al., 2013). Different rates and combination of fertilizer sources have shown good results (Kayeke et al., 2013), whereas sources of N, amount applied, and time of application are very important to achieve effective weed management (Parker and Riches, 1993; Sauerborn, 1991).

6.3.7 Mulching/Ground Cover

The smothering effect of mulch on weeds has long been considered an important method of reducing weed densities. Both live and dead mulch have been reported to assist crop establishment and development. However, where crop–livestock farming is dominant, mulching with crop residues has created competition for use of the resource. Mulch has multiple ways in which it affects weed vigor, including physical interference, allelopathy on some weed species, and limiting access to light. On the other hand, mulching provides conducive soil environment for weed seed germination in the form of soil moisture and temperature for species.

6.4 WEED MANAGEMENT IN SMALLHOLDER CROPPING SYSTEMS

Systems research has taken center stage in recent years with the view of improving and delivering both agricultural products and at the same time ecological services related to cropping systems. In weed research systems analysis has assisted the development of plot level hypothesis to fit into an ecosystems application of the scientific logic. The transition from the plot level evaluation has improved the systems understanding of interaction leading to the inclusion of more factors and permutations in the analysis of the outcomes. Save for proper timing of weeding operations there are no general rules promoted yet for managing weeds in different cropping system.

Crop combinations and sequences are a common practice in southern Africa. Work by Liebman and Dyck (1993) revealed that a significant reduction of weed population and biomass was achieved through the use of both crop rotation and intercropping. However, the effectiveness of the temporal diversification exhibited in crop rotation and spatial diversification in intercropping strategies has not been optimized by smallholder farmers. Besides the commonalities in cropping systems described above, cropping systems practiced in similar ecologies also can utilize similar weed management strategies, e.g., upland crops or lowland cropping system. To this end there are four dominant cropping systems identified in southern Africa, namely: cassava-, maize-, rice-, and sorghum-based cropping systems that warrant focusing attention with regards to weed management practices. Table 6.2 shows some of the common weeds on arable land on smallholder farms. We hypothesize that mixed cropping systems will benefit systems productivity and also minimize weed interference under variable climate in southern Africa.

6.4.1 Cassava-Based Cropping System

Cassava is a long duration crop and varieties grown in the region take 12–18 months to maturity and these are commonly grown as monocropping and multiple cropping systems in Angola, Malawi, Mozambique,

Table 6.2 Weed species distribution from arable land in Zimbabwe surveyed immediately after crop harvest under maize-based cropping system

Treatment	Contribution of species to sample mean (%)	Critical weed characteristics
Chromolaena odorata (L.) R.M. King and H. Robinson	0.10	Perennial and spread by seed; weed is both competitive and allelopathic
Cissampelos mucronata	0.01	Annual, climbing, and spread by rhizomes
Commelina benghalensis	0.07	Annual/perennial, spread by seed and vegetative
Conyza bonariensis (L.) Cronq.	20.6	Perennial, erect, branching at base and seed propagated
Cynodon dactylon (L.) pers.	0.03	Perennial, flattened leaf blade sharp tip and spread by rhizomes
Helichrysum argyrophyllum (DC)	0.11	Perennial and propagation by seed and cuttings
Helichrysum spp.	17.65	Annual/perennial and spread by seed
Indigofera vicioides	0.01	
Oldenlandia herbacea	0.01	Annual/perennial, small leaves, and spread by seed
Richardia scabra (L.)	25.8	Perennial, broadleaf, and seed propagated
Other	0.02	

Source: Nhamo, N., 2007. The contribution of different fauna communities to improved soil health: a case of Zimbabwean soils under conservation agriculture. PhD thesis, Rheinischen Friedrich-Wilhelms-Universitat, Bonn.

Zambia, and Zimbabwe. The crop stays on the field three to four times longer than cereals such as maize, rice, and sorghum. The growth pattern presents both opportunities and threats from climate variability in that the chance of occurrence of an extreme weather event increases over the long growth period. On the other hand, the likelihood of recovery from the effects of extreme events during the crop cycle is higher compared to short duration crops. Two characteristics, which dominate cassava systems and could be major consideration when designing weed management strategies, are the crop cycle, i.e., time to maturity, and the crop architecture. Both the branching and nonbranching cultivars are widely grown in southern Africa. However, within these categories the rates with which canopies establish vary. Weed management in cassava cropping systems is currently dominated by mechanical methods with manual hand weeding leading other methods. However, with the impending cassava intensification other options that include combinations of mechanical and chemical weeding are highly likely to be effective ways of keeping the weed interference effects low.

Herbicide use in cassava is low, a trend that cuts across all the common root and tuber crops. Health concerns related to the direct consumption of cassava roots and pesticide residues need to be dispelled using recent qualitative laboratory analytical data. Even more challenging is the development of low residue herbicide formulations specifically for root and tuber crops that farmers directly consume at harvest. Food safety issues will dominate the future discussion on the use of chemical weed control in cassava systems. Nevertheless, work on a suite of weed management techniques tailor made for long duration crops, such as cassava, is needed urgently.

6.4.2 Maize-Based Cropping System

Maize systems are widely studied in southern Africa and the relative cost of weed management per unit produce and area can be calculated. Large gaps in the cost of weeding still exist because smallholder farmers largely use manual weeding methods, which are slow, labor intensive, and inefficient in southern Africa.

Use of herbicides and other external inputs such as fertilizers on hybrid maize varieties has been widely accepted. However, the herbicide value chain still has rogue elements where fake chemicals are still sold to farmers leading to lack of trust and financial losses on the part of farmers. Lack of experience and training on handling herbicides has been identified as a major hindrance to the use of chemical weed control practices on smallholder farms.

6.4.3 Rice-Based Cropping System

Rice production has a niche in the lowlands of southern Africa. Export rice brands are commonly traded on the international market and notable is rice from Malawi and Zambia. Weed management in rice cropping system is tricky as rice ecology varies from rainfed, hydromorphic to flooded/irrigated areas. In these three ecologies, there are varieties of weed species, both annual and perennial, including grass, broadleaf, and sedges. Furthermore, the parasitic weeds of Orobanchaceae family, *S. asiatica, Striga aspera, S. hermonthica* and *R. fistulosa*, can be found in these ecologies. *Striga* species are found in upland rainfed ecology, whereas *R. fistulosa* is found in hydromorphic ecology. Other weed species like *O. longistaminata* can be found in all ecologies. Weed management in these ecologies differs in relation to the type of machinery used for mechanical weed management, some methods, e.g., mulching or rotational, are not applicable to some ecologies. There are some cases where chemical formulations differ from ecology to ecology like Oxidation 250 EC used in upland and hydromorphic, whereas Oxidation 12 PL is used in flooded/irrigated ecology. Therefore, the development of integrated weed management needs a focus on weed ecology, species, and crop varieties especially those that have weed resistance characteristics in general and parasitic weed resistance in particular (Rodenburg and Johnson, 2009).

6.4.4 Sorghum-Based Cropping System

Sorghum is grown in drylands in all countries in southern Africa. It is the stable source of starch in Botswana and Namibia including drier parts of all countries in southern Africa. Weed management in sorghum has also benefited from the intercropping, relay, and strip cropping, which are common practices in sorghum systems. Sorghum is often grown in drier ecologies and dominant weed species in these ecologies have drought resilience. However, the frequency of weeding under harsh conditions is limited more by the drought. Under these conditions, the timing is highly critical as there are only a few opportunities where weed management practices can be effectively applied. Further climate change under the dry regions will probably reduce overall biomass production and effect of weeds on the yields may not be as significant. There is a need for more data on the development of weed species in drier areas compared to more humid zones and the interactive effects due to competition for moisture.

6.5 YIELD GAINS FROM APPROPRIATE WEED MANAGEMENT PRACTICES

The work of Nhamo et al. (2014) highlighted the importance of weed management practice and their contribution to yield gains in rice systems. Proper weed management practices rank first among other agronomic practices in terms of contributing to improved grain yield. The effectiveness of weed management practices cannot be valued more than the yield benefit it results in. Appropriate weed management has a large benefit to the crops under cultivation through reduced cost of production, reduced risk of contamination of produce, and reduced contribution to the weed seed bank.

6.6 RESEARCH GAPS AND NEW APPROACHES

Climate change has direct and indirect effects on the crops, more work is required to enable farmers to cope and adapt to the climate changes. Various research activities (Kimenye, 2014) have been conducted and some are underway but there are a number of research gaps on knowledge surrounding various areas of specialization. Farmers will cope and adapt climate change effects on their cropping systems as they access improved technologies.

6.6.1 Climate Change Scenarios

Climate change effects vary with location and time (years). These changes have adverse effects on ecology and biology of crops and weeds in different ways, more research work is required to evaluate the magnitude and impact of these effects and the available adaptation measures against climate variability. Research aiming at adaptation to climate variability through breeding for drought-tolerant and early maturing crop varieties (Kimenye, 2014) without consideration of the weed problem in the cropping system; a gap that needs to be addressed. When weed management research is addressed the focus should be on weed-resistant crop varieties, for example, against parasitic weeds, and very little has been done on crop varieties with robust growth characteristics to overcome weeds in the field, e.g., the New Rice for Africa varieties (Rodenburg and Johnson, 2009). Another research gap to be addressed is on the dynamics of weed species in different ecologies and related cropping systems under different scenarios of climate variability.

6.6.2 Weed Biology and Ecology

Climate variability can result in erratic, increased, or reduced rainfall; this can change the ecology of a particular area from dry land to hydromorphic conditions and vice versa. Under this situation, weed species that can dwell in one ecology shift to another ecology. The scientific information of the new ecology will be required to develop appropriate integrated weed management strategy. In the new ecology, weed association with other plants or crops, microfauna, and macrofauna needs to be studied as it has a direct and indirect effect on the growth and development of the crop and weed species, for example, the larva of *Smicronyx* spp., a beetle feeding on capsules of *R. fistulosa* weeds observed in Kyela, Tanzania and African gall midge life cycle in *O. longistaminata* in lowland flooded rice ecology. Another research gap is the need to understand weed dynamics across more than one ecology and potential extinction as the ecology changes, for example, aquatic weeds *Salvinia* spp. cannot survive in dry conditions. Furthermore, studies must be carried for some weeds with scant scientific information on their biology and ecologies such as *R. fistulosa* (Kayeke et al., 2010; Rodenburg, et al. 2014) to develop the needed information to narrow the knowledge gap. Developed scientific information is needed as baseline information for the development of integrated weed management strategy (Rodenburg et al., 2014).

6.6.3 Development of Herbicide Resistance

Herbicide resistance is the ability of a weed species to avoid the effect of the chemical weed control applied (Osuna et al., 2011). There are two main mechanisms of resistance: nontarget–site resistance and target–site resistance as reported by Powles and Yu (2010). According to Powles and Yu (2010), non-target–site resistance is the alteration of the nucleotide sequences of the gene to which the herbicide binds and prevents or stops herbicide action, whereas target–site resistance occurs as a result of mutation or gene duplication. Research is needed to reduce the adverse effect by the development of models that can predict scenarios that can increase chances of herbicide resistance as an indicator of effect of climate to crops. Identification of naturally available weed resistance in weed population in the cropping system remains an important research gap. This will enable the deployment of specific measure to contain specific species in their natural geographical area as the spread to other cultivated area can magnify the problem (Chali, 2015). Another area of interest for research is where there is gene flow to and from crop and the close wild relative species as observed in rice, e.g., *Oryza sativa* and *O. longistaminata* (Kilewa, 2014). The information on gene flow is very important especially when there are efforts to develop herbicide-resistant crop varieties to avoid promotion of weediness, difficulty in weed management, and extinction of wild species, which are sources of useful traits in breeding programs. On the other hand, the research should focus on the development of an integrated weed management strategy, which will reduce the chances of development of herbicide resistance among weed species either as a result of application of chemical control or through natural selection.

6.6.4 Development of Climate Smart Weed Management Technology

Currently, there are some research efforts on climate smart agriculture and specifically on weed management technology. These include a push pull technology, which is an intercropping technology where maize and *Desmodium uncinatum* are boarded by Napier grass (*Pennisetum purpureum*) (Cook et al., 2006; Khan et al., 2010; Midega et al., 2009). Much is needed to address weed management in different ecologies for different crops under different climate change scenarios. Furthermore, much is needed for the research to develop climate smart weed management technologies as components of integrated weed management. In view of cultural weed management such as cover crops in rotation or intercropping, there is a need for research to analyze the sustainability of the cropping system (Hayati et al., 2010). This will inform how long the system will be stable, profitably and sustainable systems. The multidisciplinary analysis suggested covers more than one aspect of weed management, e.g., ecology and economics of practices and value (Hayati et al., 2010; Van Passel, 2008).

6.6.5 Analysis of Plant Protection System

As inefficiency of the biosecurity system can result in poorly addressed problems of weed management in a particular country. Research must assess plant protection or biosecurity system of a given geographical area as reported by (Schut et al. 2015a,b). This will involve institutional analysis at various scales, i.e., regional level to farm level in order to identify weakness of the weed management systems. Furthermore, the system analysis enables determination of the magnitude of the problem, availability, and adequacy of management strategies.

6.6.6 Relevance of Weed Management to Scale of Farm Operation

Weak weed management development strategies affect the current cropping practices. Two extreme scenarios do exist, where resource-poor smallholder farmers are struggling with manual weeding (weeding takes 60% of the labor supplied to the production of crops) and potentially high use of herbicide by medium- to large-scale farmers. From these extremes, poor returns to labor and herbicide residue accumulation in both soils and water bodies are the major outcomes of concern. There is need for research to find a middle of the road solution where the negative environmental externalities and human health risks associated with overuse of chemical weeding and the huge labor burden of manually using hand-hoe and hand-pushed weeders is addressed.

6.6.7 Breeding for Suppression of Weeds

Breeding efforts, which were initiated in the 1970 after the discovery of allelopathic plants, could shed more light into how the allelopathic interference can lead to weed suppression from acceptable crop cultivars (Belz, 2007).

6.6.8 Combined Weed Management Practices

Judicious use of mixed mechanical and chemical weeding on specific enterprises increases efficiency and also cost effectiveness in an environmental way. Climate change will enhance weed species that are resistant and tolerant to extremes in a similar way, some species have begun to show herbicide tolerance. More work on integrated weed management is required, knowledge on the evolution of weed species and their response to herbicides need to be combined with that of mechanical and biological approaches.

6.6.9 Interference in Crop–Weed Relations

Weed competition is dependent on both the crop factors and the weed factors; a lot of work has focused on the interference on the crop but has neglected the data on weeds. Research with an ecological focus is required to elicit the competition effect on the weed flora and also the resultant impact on the weed seed bank in the soil. Such a study could be done combining both controlled and field environments and also on climate extremes affected locations.

6.7 CONCLUSION

Weeds pose a real risk to crop production and climate extremes exacerbate this situation especially on upland ecologies where management is critical at specified times. A lot of work has been done on weed management practices but challenges still remain in the area of management of weed seed bank, taking to scale combined weed management practices, reducing the herbicide residue on root crops, and increasing awareness of the correct use of herbicides to reduce cost of weed management. Across cassava, maize, and sorghum systems, major weed management considerations will be on the crop phenology, rainfall variability, and cultivar-specific traits that challenge smart weed management practices.

REFERENCES

Akobundu, O.I., 1987. Weed Science in the Tropics Principles and Practices. John Wiley and Sons, Norwich, p. 522.

Azam-Ali, S.N., Squire, G.R., 2002. Principal of Tropical Agronomy. Biddles Ltd, Guildford and King's Lynn, UK, p. 236.

Belz, R.G., 2007. Allelopathy in crop/weed interactions—an update. Pest Management Science 63 (4), 308–326.

Bunce, J.A., 1993. Growth, survival, competition, and canopy carbondioxide and water vapor exchange of first year alfalfa at an elevated CO2 concentration. Photosynthetica 29, 557–565.

Chali, T., 2015. Assessment of Glyphosate Tolerant Weed Biotypes in Three Rice Growing Regions of Tanzania M.Sc. thesis. Adelaide University, Australia.

Charudattan, R., 1986. Integrated control of water hyacinth (*Eichhornia crassipes*) with a pathogen, insects, and herbicides. Weed Science 34, 26–30.

Chauhan, B.S., Gill, G.S., 2014. Ecologically based weed management strategies. In: Chauhan, B.S., Mahajan, G. (Eds.), Recent Advances in Weed Management. Springer, New York, pp. 1–11.

Chikwenhere, G.P., Keswani, C.L., 1997. Economics of biological control of Kariba weed (*Salvinia molesta* Mitchell) at Tengwe in north-western Zimbabwe-a case study. International Journal Pest Management 43 (2), 109–112.

Clements, R.D., DiTommaso, A., Hyvönen, T., 2014. Ecology and management of weeds in a changing climate. In: Chauhan, B.S., Mahajan, G. (Eds.), Recent Advances in Weed Management. Springer, New York.

Cook, S.M., Khan, Z.R., Pickett, J.A., 2006. The use of push-pull strategies in integrated pest management. Annual Review of Entomology 52 (1), 375.

Emechebe, A.M., Ellis-Jones, J., Schulz, S., Chikoye, D., Douthwaite, B., Kureh, I., Tarawali, G., Hussaini, M.A., Kormawa, P., Sanni, A., 2004. Farmers' perception of the *Striga* problem and its control in Northern Nigeria. Experimental Agriculture 40 (02), 215–232.

Gitay, H., Suarez, A., Watson, R.T., Dokken, D.J., 2002. Climate change and biodiversity. In: Technical Paper of the Intergovernmental Panel on Climate Change. IPCC Secretariat, Geneva, p. 77.

Hayati, D., Ranjbar, Z., Karami, E., 2010. Measuring agricultural sustainability. In: Lichtfouse, E. (Ed.), Biodiversity, Biofuels, Agroforestry and Conservation AgricultureSustainable Agriculture Reviews, 5, p. 73. http://dx.doi.org/10.1007/978-90-481-9513-8_2.

Jablonkai, I., 2015. Herbicide Metabolism in Weeds—Selectivity and Herbicide Resistance. http://dx.doi.org/10.5772/61674.

Kabiri, S., Rodenburg, J., Kayeke, J., Van Ast, A., Makokha, D.W., Msangi, S.H., Irakiza, R., Bastiaans, L., 2015. Can the parasitic weeds *Striga asiatica* and *Rhamphicarpa fistulosa* co-occur in rain-fed rice? Weed Research 55. http://dx.doi.org/10.1111/wre.12124.

Kamara, A.Y., Chikoye, D., Ekeleme, F., Omoigui, L.O., Dugje, I.Y., 2008. Field performance of improved cowpea varieties under conditions of natural infestation by the parasitic weed *Striga gesnerioides*. International Journal Pest Management 54 (3), 189–195.

Kayeke, J., Sibuga, P.K., Msaky, J.J., Mbwaga, A., 2007. Green manure and inorganic fertilizer as management strategy for witch-weed in upland rice. African Crop Science Journal 15 (4), 161–171.

Kayeke, J., Rodenburg, J., Mwalyego, F., Mghogho, R., 2010. Incidence and severity of the facultative parasitic weed *Rhamphicarpa fistulosa* in lowland rainfed rice in southern Tanzania. In: Second Africa Rice Congress, Bamako, March 22–26, 2010.

Kayeke, J., Rodenburg, J., Bastiaans, L., Menza, M.K., Onyuka, E.A., Cissoko, C., Daniel, E.D., Makokha, D.W., Van Ast, A., 2013. Locally adapted parasitic weed management strategies based on soil fertility amendments. In: Third Africa Rice Congress, October 21–24, 2013, Yaoundé, Cameroon.

Khan, Z.R., Midega, C.A., Bruce, T.J., Hooper, A.M., Pickett, J.A., 2010. Exploiting phyto-chemicals for developing a "push–pull" crop protection strategy for cereal farmers in Africa. Journal of Experimental Botany 61 (15), 4185–4196.

Khan, Z., Midega, C., Pittchar, J., Pickett, J., Bruce, T., 2011. Push–pull technology: a conservation agriculture approach for integrated management of insect pests, weeds and soil health in Africa: UK government's Foresight Food and Farming Futures project. International Journal of Agricultural Sustainability 9 (1), 162–170.

Kilewa, R., 2014. Distribution of wild rice species and hybridization between cultivated rice and Oryza longistaminata M.sc. thesis. Adelaide University, Australia.

Kimenye, L. (Ed.), 2014. Best-bet Technologies for Addressing Climate Change and Variability in Eastern and Central Africa. ASARECA (Association for Strengthening Agricultural Research in Eastern and Central Africa), Entebbe, p. 224.

Liebman, M., Dyck, E., 1993. Crop rotation and intercropping strategies for weed management. Ecological Applications 3 (1), 92–122.

Masifwa, W.F., Twongo, T., Denny, P., 2001. The impact of water hyacinth, Eichhornia crassipes (Mart) Solms on the abundance and diversity of aquatic macroinvertebrates along the shores of northern Lake Victoria, Uganda. Hydrobiologia 452 (1–3), 79–88.

Mengel, K., Kirkby, E.A., 1987. Principal of Plant Nutrition. International Potash Institute, Bern, Switzerland, p. 685.

Midega, C.A., Khan, Z.R., Van den Berg, J., Ogol, C.K., Bruce, T.J., Pickett, J.A., 2009. Non-target effects of the "push–pull" habitat management strategy: parasitoid activity and soil fauna abundance. Crop Protection 28 (12), 1045–1051.

Mitchell, D.S., 1972. The kariba weed: Salvinia molesta. The British Fern Gazette 10 (5), 251–252.

Nhamo, N., 2007. The contribution of different fauna communities to improved soil health: a case of Zimbabwean soils under conservation agriculture PhD thesis. Rheinischen Friedrich-Wilhelms-Universitat, Bonn.

Nhamo, N., Rodenburg, J., Zenna, N., Makombe, G., Luzi-Kihupi, A., 2014. Narrowing the rice yield gap in East and Southern Africa: using and adapting existing technologies. Agricultural Systems 131, 45–55.

Osuna, M.D., Okada, M., Ahmad, R., Fischer, A.J., Jasieniuk, M., 2011. Genetic diversity and spread of thiobencarb resistant early watergrass (Echinochloa oryzoides) in California. Weed Science 59 (2), 195–201.

Parker, C., Riches, C.R., 1993. Parasitic Weeds of the World: Biology and Control. CAB International, Wallingford, UK. 332 pp.

Peters, K., Breitsameter, L., Gerowitt, B., 2014. Impact of climate change on weeds in agriculture: a review. Agronomy for Sustainable Development 34 (4), 707–721.

Pickett, J.A., Woodcock, C.M., Midega, C.A., Khan, Z.R., 2014. Push–pull farming systems. Current Opinion in Biotechnology 26, 125–132.

Powles, S.B., Yu, Q., 2010. Evolution in action: plants resistant to herbicides. Annual Review of Plant Biology 61, 317–347.

Rodenburg, J., Johnson, D.E., 2009. Weed management in rice-based cropping systems in Africa. Advances in Agronomy 103, 149–218.

Rodenburg, J., Johnson, D.E., 2013. Managing weeds of rice in Africa. In: Wopereis, M.C.S., Johnson, D.E., Ahmadi, N., Tollens, E., Jalloh, A. (Eds.), Realizing Africa's Rice Promise. CABI, Wallingford, Oxfordshire, UK, pp. 35–45.

Rodenburg, J., Morawetz, J., Bastiaans, L., 2014. Rhamphicarpa fistulosa, a widespread facultative hemi-parasitic weed, threatening rice production in Africa. Weed Research. http://dx.doi.org/10.1111/wre.12129.

Rodenburg, J., Cissoko, M., Kayeke, J., Dieng, I., Khan, Z.R., Midega, C.A.O., Onyuka, E.A., Scholes, J.D., 2015. Do NERICA rice cultivars express resistance to Striga hermonthica

(Del.) Benth. and *Striga asiatica* (L.) Kuntze under field conditions? Field Crops Research 170, 83–94.

Rosenzweig, C., Iglesias, A., Yang, X.B., Epstein, P.R., Chivian, E., 2001. Climate change and extreme weather events; implications for food production, plant diseases, and pests. Global Change & Human Health 2 (2), 90–104.

Sauerborn, J., 1991. Parasitic Flowering Plants. Ecology and Management. Verlag Josef Margraf, Weiksheim. 127 pp.

Schut, M., Rodenburg, J., Klerkx, L., Hinnou, L.C., Kayeke, J., Bastiaans, L., 2015a. Participatory appraisal of institutional and political constraints and opportunities for innovation to address parasitic weeds in rice. Crop Protection 74, 158–170.

Schut, M., Rodenburg, J., Klerkx, L., Kayeke, J., Van Ast, A., Bastiaans, L., 2015b. RAAIS: rapid appraisal of agricultural innovation systems (part II). RAAIS: rapid appraisal of agricultural innovation systems (part II). Integrated analysis of parasitic weed problems in rice in Tanzania. Agricultural Systems 132, 12–24.

Van Passel, S., 2008. Assessing farm sustainability with value oriented methods. In: 12th Congress of the European Association of Agricultural Economists – EAAE, pp. 1–5.

Xuan, T.D., Shinkichi, T., Hong, N.H., Khanh, T.D., Min, C.I., 2004. Assessment of phytotoxic action of *Ageratum conyzoides* L. (billy goat weed) on weeds. Crop Protection 23 (10), 915–922.

Ziska, L.H., Teasdale, J.R., Bunce, J.A., 1999. Future atmospheric carbon dioxide may increase tolerance to glyphosate. Weed Science 47, 608–615.

FURTHER READING

Liebman, M., Mohler, C.L., Staver, C.P., 2001. Ecological Management of Agricultural Weeds. Cambridge University Press.

Rodenburg, J., Riches, C.R., Kayeke, J.M., 2010. Addressing current and future problems of parasitic weeds in rice. Crop Protection 29, 210–221.

Rodenburg, J., Kayeke, J., Van Ast, A., Bastiaans, L., 2013. The potential of timing as a parasitic weed management strategy for smallholder rice farmers. In: 12th World Congress on Parasitic Plants, July 15–19, 2013, Sheffield, UK.

Satrapová1, J., Hyvönen, T., Venclová1, V., Soukup, J., 2013. Growth and reproductive characteristics of C4 weeds under climatic conditions of the Czech Republic. Plant Soil Environment 59 (7), 309–315.

Sun, W., Li, Q., Fan, Y., Wan, Y., Wang, T., Cong, B., 2015. Effect factor analysis of spraying quality for agricultural chemicals. International Journal of u- and e- Service, Science and Technology 8 (11), 221–230.

Tu, M., Hurd, C., Randall, J.M., 2001. Weed Control Methods Handbook. The Nature Conservancy. Version: April 2001. http://tncweeds.ucdavis.edu.

Varanasi, A., VaraPrasad, P.V., Jugulam, M., 2015. Impact of climate change factors on weeds and herbicide efficacy. Advances in Agronomy. http://dx.doi.org/10.1016/bs.agron.

Vencill, W.K., Nichols, R.L., Webster, T.M., Soteres, J.K., Mallory-Smith, C., Burgos, N.R., Johnson, W.G., McClelland, M.R., 2012. Herbicide resistance: toward an understanding of resistance development and the impact of herbicide-resistant crops. Weed Science, (special issue: 2–30).

Yu, Q., Cairns, A., Powles, S., 2007. Glyphosate, paraquat and ACCase multiple herbicide resistance evolved in a *Lolium rigidum* biotype. Planta 225 (2), 499–513.

CHAPTER 7

Opportunities for Smallholder Farmers to Benefit From Conservation Agricultural Practices

Nhamo Nhamo[1], Olipa N. Lungu[2]

[1]International Institute of Tropical Agriculture (IITA), Southern Africa Research and Administration Hub (SARAH) Campus, Lusaka, Zambia; [2]University of Zambia (UNZA), Lusaka, Zambia

Contents

7.1 INTRODUCTION

In southern Africa, the popularity of conservation agriculture (CA) is still growing, with support from government agencies, nongovernmental organizations (NGOs), the Food and Agriculture Organization of the United Nations (FAO), and numerous development partners (Nyamangara et al., 2013a,b). Similar promotion efforts, leading to increased adoption trends,

Smart Technologies for Sustainable Smallholder Agriculture
ISBN 978-0-12-810521-4
http://dx.doi.org/10.1016/B978-0-12-810521-4.00007-4
145

have been reported in North and South America (Derpsch, 2008; Holland, 2004; FAO, 2010). CA has been viewed as a set of high potential and sustainable agricultural practices that could contribute to ecological resilience and climate change adaptation and mitigation, and is believed to be suitable for low-income smallholder farmers (Thierfelder et al., 2013). In most cases, promotion efforts on CA have been targeting the poor smallholder farmers in southern Africa.

Despite the positive views on CA, a substantial amount of questions has been recently raised on (1) the definition and what constitutes CA, (2) which environments CA is most suitable for, (3) the economics of the practices on smallholder farms, (4) the practical application of mulching in crop–livestock systems, and (5) which farmer typology the technologies should target. Questions have also been raised on the ecological contribution of CA to food production systems in relation to the scale of operation under smallholder farming systems; typical smallholder farmers cultivate fragmented portions of land distributed around their village boundaries (Blarel et al., 1992). The current scenario is such that the total area under CA on smallholder farms has probably not significantly increased (Baudron et al., 2007) to warrant resounding ecological benefits across the region; South Africa, Tanzania, Democratic Republic of Congo, and Mozambique are still warming up to the massive scaling out of CA. Ngwira et al. (2014) have shown increased rates of adoption of CA in Malawi albeit on small pieces of land, e.g., less than 2.2 ha. In Zimbabwe, where uptake of CA was incentivized by agricultural inputs, adoption rates on individual farms seldom go beyond plots larger than 0.2 ha for which the inputs were provided by humanitarian organizations (Anderson and D'Souza, 2014). However, there are also doubts over CA's large-scale effects on greenhouse gas emissions, effectiveness on variable rainfall patterns, and temperature extremes in densely populated areas and where agricultural land holdings are reduced drastically. There is therefore need to examine the constraints limiting CA application to larger areas and the opportunities for improvements in the implementation in countries where potential benefits are high.

CA is primarily based on three principles: (1) soil surface cover using dead or live organic mulching materials, (2) minimum soil disturbance with reduced tillage, and (3) crop combinations in the form of rotations and associations of different crop cultivars (FAO, 2010; Hobbs, 2007; IIR and ACT, 2005; Scopel et al., 2013). CA in Malawi, Zambia, and Zimbabwe was preceded by conservation tillage; the no-till conservation tillage (CONTIL) programs supported by the German Technical Cooperation Agency (GTZ) during the 1980s into the 1990s, which focused on tillage implements and surface structural management issues including but not limited to (tied) ridges, water harvesting pits, and contour ridge development (Vogel, 1993; Twomlow, 2000). The benefits of water harvesting techniques on crop production have been well documented. However, major bottlenecks associated with the investment in new equipment and high labor requirements seem to have offset the momentum and actual progress on the ground (Motsi et al., 2004; Mugabe, 2004; Mupangwa et al., 2006; Mutekwa and Kusangaya, 2007). Contour ridge system labor demand of typically 20 man-labor-days for construction, disposal of water rather than retention, and reduction in the amount of land available for production (at least by 15%) were viewed negatively and thus maintenance of structures was generally poor (Hagmann, 1994, 1996). In response to these constraints, other mechanical structures have been introduced on mulched CA plots and have gained prominence since then.

For the successful application of CA, other support considerations and agronomic practices have been identified and suggested. Scopel et al. (2013) provided a detailed outline on the relevance of the three CA principles and an additional two supportive practices. Similarly, the suggestion by Sommer et al. (2014) that fertilizer use should be considered the fourth CA principle implies that applying the three principles alone cannot achieve the desired results. In Zambia, for instance, the Conservation Farming Unit (CFU) has promoted CA with trees, a concept where agroforestry tree species have been incorporated for the sole reason of increasing plant nutrient sources on CA plots (CFU, 2006; Umar et al., 2011). Among the most commonly incorporated trees are the leguminous *Faidherbia albida* and *Sesbania sesban* species. Furthermore, extension services and rainfall variability were found to be the strongest determinants of CA adoption by smallholder farmers in Zambia (Arslan et al., 2014). Baudron et al. (2012) also reported the important role of adequate fertilization, timely planting, and crop protection in increasing the productivity under CA system in semiarid areas. As the region prepares for increased dissemination of climate smart agricultural technologies, the supporting biophysical and socioeconomic practices to the three CA pillars and related smart technologies need to be discussed further.

Across southern Africa, CA has been promoted widely in Malawi, Zambia, and Zimbabwe, with Zambia having introduced CA much earlier than the rest. In Zambia, increased focus on promoting CA led to the formation and development of several associations and government programs, and to date, farmer associations have continued to strongly promote CA (CFU, 2006). Despite the spread of CA in the maize growing areas in southern Africa, three observations stand out: (1) the practice remained on relatively small plots, similar to the demonstration plots used by extension personnel; (2) maize stover is the most common form of mulching material; and (3) most farmers have adopted a strict set of annual crop combinations on CA plots to the detriment of the diversity of CA systems. However, the CFU in Zambia has promoted CA on maize systems, which included agroforestry species (annual and perennial) an innovation different from other countries (CFU, 2006; Nyamangara et al., 2013a,b). Furthermore, maize production in southern Africa is supported by government programs: the Farmer Input Support Program (FISP) in Zambia, the Agricultural Input Subsidy Program (AISP) in Malawi, and similar programs in Tanzania and Zimbabwe, which provide subsidized fertilizer and seed inputs (Lunduka et al., 2013; Jayne and Rashid, 2013), and often the fertilizer is also applied on the CA plots. Despite this large-scale support, only a small section of farmers outside areas targeted by projects such as CFU (<5%) have adopted some form of CA in Zambia (Arslan et al., 2014).

Although there is evidence of use and popularity of components of CA across southern Africa (FAO, 2010; Thierfelder et al., 2013), especially in areas where projects have been conducted, its large-scale application in smallholder farming communities needs to be examined carefully (Giller et al., 2009; Guto et al., 2011). The aim of this chapter is to evaluate the potential models that are applicable to the widespread adoption of CA by smallholder farmers in southern Africa. The chapter has been structured into the following: Section 7.2, CA Strengths and Weaknesses; Section 7.3, Minimum Investment Requirements for CA systems; Section 7.4, Ecological Indicators of CA; Section 7.5, Research Gaps; and Section 7.6, Conclusions.

7.2 CA STRENGTHS AND WEAKNESSES

The practices of CA in southern Africa have developed based on different definitions and this results in the application of different packages from country to country (Anderson and D'Souza, 2014). The definitions have differed in the form of physical structures in use, with hand-hoe-made basins for soil moisture conservation being the most popular. The size of the basins varies; however, the $15\,cm \times 15\,cm \times 15\,cm$ was reported most frequently. Among the equipment required for CA, e.g., rippers, the Magoye ripper was the most adapted as the tine could be fitted on a moldboard plow frame. Direct seeding equipment that are hand driven, draught power drawn, and tractor operated have been used whereas the Brazilian animal-drawn direct seeder (Fitarelli No. 12) is being promoted in Malawi, Zambia, and Zimbabwe; the seeder has the advantage of combining tillage, seed, and fertilizer application in one frame. Owing to the differences in the investment requirements, CA has evolved differently across countries in southern Africa. Nonetheless, some benefits have been reported in the literature from CA, which can be extrapolated to similar environments; at the same time, some weaknesses that have constrained adoption of practices by farmers have also surfaced.

7.2.1 Benefits of CA

The benefits of CA vary, and improved maize yield (productivity), infiltration and soil moisture, soil organic carbon (SOC; physicochemical factors), and reduced weed pressure, pests, and diseases (biotic factors) have been discussed widely. Increase in yield following the use of CA practices was the major benefit. Thierfelder et al. (2013) reported yield gains of 5 t/ha on maize grain and 1 t/ha grain on cowpea following the application of CA in Zimbabwe and Zambia. This milestone for smallholder farming systems was significant, given that the average yield for maize has stagnated around 0.89 t/ha for the past two decades; rate of population increase exceeds increase in food production and land holding sizes decreased. However, it has been difficult to quantify and separate yield gains resulting from the components of CA, i.e., soil cover, crop combinations, and reduced tillage, and from those emanating from other supporting practices, e.g., access to fertilizer and extension intensity during the contact period (Arslan et al., 2014; Sommer et al., 2014). Giller et al. (2015) further argued that the observed yield gains were mostly as a result of better management practices, i.e., proper timing of sowing, early crop establishment, and effective weed management. Additionally, the benefits of better management of smaller pieces of land by small groups of cooperating farmers (which is a challenge for the poorest smallholder farmers) have contributed to increased yields obtained using CA. In view of high crop yield obtained using a range of CA practices, a systems agronomy approach has been recommended, which can link CA practices to sustainable intensification and can provide a thorough analysis of the issues surrounding yield gains (Giller et al., 2015; Vanlauwe et al., 2014).

In situ retention of crop residues coupled with no-till practices have shown great potential to increase SOC content in the long run. The effect of these CA practices on soil organic matter (SOM) dynamics have been reported widely (Devkota et al., 2015a,b; Ibragimov et al., 2011; Djalankuzov and Saparov, 2009; Egamberdiev, 2007; Lal, 2015; Tursunov, 2009; Pulatov et al., 2012). The organic matter from the mulch is known to contribute to the promotion of more stable soil aggregates as a result of increased microbial activity and better protection of the soil surface (Hobbs, 2007). However, reports on the lack of significant changes in SOM in the soils after applying CA practices (Nyamangara et al., 2013a,b) suggest that building up of SOM does not apply to all soil types. Higher SOM levels in soils under CA result in more biotic diversity support because of the influence of the mulch and minimum disturbance of nesting sites, a significant soil ecological benefit.

Soil erosion directly affects soil physical properties and subsequently soil quality, which is an important attribute for sustainable cropping. Reduced tillage and mulching have the overall effect of reducing the effect of both wind and water erosion. In southern Africa, where soils are relatively sandy with limited clay content and SOM, excessive erosion will reduce the potential of the soil to support crop production. Soil erosion, a major indicator of soil degradation from arable land, has been estimated at 50 t/ha year (Elwell and Stocking, 1988). Elwell (1986) related erodibility of these soils with land use and management. Protection of the soil through CA practices such as reduced soil disturbance and soil cover has great potential to reduce the chances of erosion of bare soil.

One of the major drivers of climate change is the anthropogenic activities such as agricultural productivity. This has continued to contribute a great deal to global warming through the production of greenhouse gases (methane, nitrogen oxide, and carbon dioxide). Increased carbon dioxide concentration in the atmosphere has a very huge impact on ozone production and subsequently global warming. CA has been reported in a number of studies in temperate climates to have the capacity to sequester soil carbon. The increase in SOC under CA directly contributes to reduction in emissions leading to the mitigation of climate change, as more carbon is stored in the soil. In this process of carbon sequestration, the soil becomes a very important sink of carbon, thereby contributing to the reduction of greenhouse gas levels. However, studies in sub-Saharan Africa have not explicitly shown the extent to which this process is evident in CA systems. The fragmented farms in the smallholder farming systems also need to be integrated for this effect to be significant at continental and world levels.

The practice of applying limited tillage and hence reduced exposure of soil carbon and the protective effect of mulching are the major drivers of this mitigation effect under CA. However, this effect will be short lived in instances where farmers revert to conventional plowing again. For sustainability, long-term application of CA would be required especially for coarse-textured soils. Limiting tillage has its own negative effects on the soil profile development albeit its important effect in CA systems. Therefore, there is need for farmers to employ the recommended use of rippers to correct ponding, crusting, and other physical challenges that may result from applying CA over a long time. At present, only a few farmers have invested in additional equipment because they are costly, e.g., rippers. Management of carbon dynamics under CA is such a critical step in defining CA as a climate smart technology, as it affects carbon sequestration and SOM buildup, and hence, smallholder farmers should be made aware of this.

Apart from reducing soil moisture content, climate extremes have the capacity to reduce the population and diversity of soil organisms involved in important soil processes, which support environmental services and soil systems development. According to Classen et al. (2015), microbes in the soil are affected in different ways and sometimes change their morphology and function to adapt to extreme climate events. This impacts the soil health as a whole, at levels depending on the extent of the modification and severity on the population affected. On the other hand, with reduced tillage and soil cover, the nesting sites for soil organisms increase in CA systems, and a study by Kladviko (2001) with particular interests in earthworms revealed that CA systems favored biota in soil. The development of a food web under reduced tillage systems has often been discussed in the context of soil health. Although soil health goes beyond the mere numbers of organisms and the trophic levels of a food web, supporting reduced tillage systems with organic matter addition is based on the hypothesis that a balanced food web is a precursor to healthy soil (FAO, 2005). Nhamo et al. (2005) have shown movement of termite densities in response to both management of organic matter and also niche characteristics, e.g., rainfall and temperature. Fig. 7.1 shows the variation of beetle larva following the application of different CA systems in Zimbabwe. More data has shown the population dynamics of soil fauna under conventional plowing systems and CA systems, with high densities supported by the latter (Nhamo, 2007).

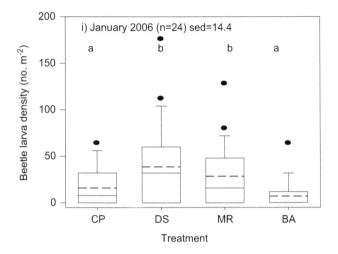

Figure 7.1 Changes in beetle larva densities in relation to conventional plowing (CP), direct seeding (DS), mulch ripping (MR), and hand-hoe-made basins (BA) on sandy soils in Zimbabwe. *Nhamo (2007)*.

7.2.2 Soil Moisture

In the light of climate change, extreme drought will destroy crops unless methods of increasing water-use efficiency are in place. Soil physical characteristics and the management of organic inputs will have a bearing on water holding capacity of soils and hence on the amount of water that can be used to produce biomass. Mulching also provides soil cover, which reduces water loss from the soil surface. However, this is contingent on the type of mulch used; most reports from Malawi, Zambia, and Zimbabwe have shown that use of maize stover is most common (Nyamagara et al., 2013). Limited studies have shown use of other mulching materials that can pack more easily, providing a closer soil–mulch contact zone, which better protects the soil and reduces excessive evaporation. Besides protecting the soil from erosion and rain–drop action, smooth covers will also contribute to moderation of soil temperature and weed control. Studies on physical and biological properties of soils in Zambia showed more stable soil aggregates (41%–45%) in directly seeded CA systems compared with the conventional system (24%). This stability promoted better water infiltration and soil water storage in the directly seeded CA systems (Thierfelder et al., 2013). With better infiltration and improved water holding capacity, water productivity is bound to increase. In drier areas, soil moisture profile improvement significantly increase crop performance and reduce drought effects.

The retention of crop residues on the soil surface ultimately reduces its surface evaporation. With this reduction in evaporation, soil can retain moisture a little longer, which the crop roots and microbes can utilize. CA also improves the soil physical conditions by improving the structure, porosity, and water retention. In the event of an agricultural drought, farmers can benefit from the retained water before their crops suffer or are damaged by water stress. Water scarcity continues to be a major challenge in most parts of the country and is expected to worsen with growing population and demand by various competing users across the region (Showers, 2002). Coupled with climate change, the problem of water scarcity is likely to cause a decline in production. Therefore, this benefit from CA is advantageous to the farmers and biota. However, in the long run (>10 years), studies have shown that CA practices tend to increase the bulk density, thereby reducing the water retention in the top 25 cm of the soil (Verhulst et al., 2010).

7.2.3 Soil Water and Nutrient Synergy

Climate variability will affect synergies between management of water and soil nutrients. Small and less frequent window of opportunities for crops to respond to available nutrients and moisture will occur during the cropping season. As such cultivars that respond to water and nutrient inputs will provide better benefits to farms, especially those practicing CA. Research on water–nutrient interactions has shown that water availability to plants without providing nutrients is not helpful just as nutrient availability without the moisture is futile. Measures to conserve water in the soil without addressing the nutrient imbalance have resulted in poor crop yields (Zougmore et al., 2000), a phenomenon explained by the law of the limiting factor. Fofana et al. (2003) compared yield from wetter plots with soil water conservation (SWC) technologies and drier plots without SWC technologies and found that crop yields were explained by improvements in the quality of soil amendments. Without an efficient nutrient management system, poor water use efficiency is inevitable. The importance of studies on synchrony of soil inputs has long been acknowledged, and addressing the soil nutrient imbalances increased water use efficiency by a factor of 3–5 (Fofana et al., 2003) and 2–3 (Bationo and Mokwunye, 1991). The synergistic effects between water and nutrient management will yield sustainable gains in situations where climate change and soil degradation threaten crop production.

7.2.4 Constraints to CA Adoption

CA is a relatively new and different approach to farming for smallholder farmers in southern Africa, but there are demonstrated benefits of adopting CA practices, with finding the suitable solution to hurdles being the most logical step. The constraints to CA adoption by smallholder farmers include trade-offs between residue retention and livestock feed in mixed crop–livestock systems, the need for crop rotations and intercropping, weeding intensity and control, labor constraints, and the lack of access to complementary inputs. Often, significant knowledge gaps and the perception of farmers that tillage is necessary to produce crops are major impediments to widespread adoption of CA. To change the mind-set of farmers, efforts to simplify the steps (without compromising the principles) are required.

Weeds have been a dominant yield-reducing factor since millennia. Reduced tillage under CA tends to lead to higher weed infestations compared to conventional plowing methods in which ploughing clears the first flash of weeds. However, there are a number of mechanical, manual, and chemical methods developed to reduce weed pressure on crops. A number of selective weed killers (herbicides) are also available on the market, which the smallholder farmer can easily access and use rather than using manual labor to clear the weeds.

Labor constraints: The availability and cost of labor play a pivotal role in the likelihood of farmers to adopt a new technology. Certain forms of CA practices (e.g., the basin systems, based on manually dug planting basin) can be more labor intensive than conventional agriculture. The advent of HIV/AIDs in most parts of southern Africa has robbed the agricultural sector of the much needed labor force, thereby causing a serious strain on the existing members of the household.

7.2.5 Lack of Access to Complementary Farm Inputs and an Enabling Environment

The effect of the intertwined challenges discussed earlier is compounded by the lack of access to complementary farm inputs, such as fertilizers, herbicides, and pesticides, which could offset some of the adversities. Smallholder farmers often do not own farm equipment beyond the basic moldboard plow, a scotch cart, and a harrow. With the advent of CA, farmers need to have the suitable equipment for the cropping system. The success of this largely depends on the resource endowment of the farmers and income sources.

For large-scale operations, tractors and other tractor-drawn equipment will be required. No-till practice uses less diesel fuel and thus results in lower carbon dioxide emissions, one of the gases responsible for global warming. In this regard, CA is climate smarter than conventional plowing systems in which plowing is followed by disc harrowing and other operations that consume a lot of fuel.

7.2.6 Yield Losses

Some studies have reported yield losses from CA treatments compared to the conventional plowing treatments. Yield losses as high as about 2 t/ha were reported by Thierfelder et al. (2013). However, none of these observed yield penalties have been satisfactorily explained in the context of variable environments where CA has been tested. Giller et al. (2015) similarly reported a 10% yield penalty from a meta-analysis on global data from CA studies. It can be argued that such case studies in which CA did not result in a better crop response are important in defining the scenarios in which the practice needs to be extended and scaled out.

7.2.7 Pests and Diseases

Climate change will affect the biological systems in various ways, with some populations being favored and some being suppressed (IPCC, 2007). Although there has not been reports on major pests and diseases following the application of CA, there is a need to build contingent measures that are useful in the future. For instance, drier conditions drive the pest attack on cowpea, leading to crop losses and increased cost of production (Karungi, et al., 2000). However, where natural enemies to crop pests find a niche, the risk of pest buildup and attack is expected to reduce (Table 7.1).

7.3 THE MINIMUM INVESTMENT REQUIREMENTS FOR CONSERVATION AGRICULTURE SYSTEMS

Practicing CA entails a number of changes in the way that smallholder agriculture is conducted, perceived, and planned. Following are the key considerations to be made for CA to work on smallholder farms:

- Knowledge of CA and agronomy: Farmers stand a better chance of success with CA when they have received training on agronomy in general and specific crop management practices and CA.
- Timing operations: There is evidence that timely execution of operation on CA plots and knowledge of technologies assist farmers adopt CA. Therefore, knowledge of farm management and application of fertilizer and manure, for example, contribute to better responses of crops in CA.

Table 7.1 Conservation agriculture principles and suggested areas of improvement to make the practices more suitable for smallholder farmers

Principles	Areas of improvement	Sources
1. Use of minimum tillage	1. Finding suitable cover crops and grain legumes	Dobermann (2004), Erenstein (2011)
2. Soil protection through mulching and dead or living organic materials	2. Reducing labor and/ or costs of weed and organic matter management	FAO (2010)
3. Use of crop rotation or intercropping (crop combinations)	3. Developing suitable tillage equipment	Giller et al. (2009)
4. Biomass production whenever possible throughout the whole year	4. Creating better links to markets (using value chain approach)	Oicha et al. (2010), Scopel et al. (2013)
5. Use of multifunctional cover crops	5. Improving suitability for crop–livestock farming systems	Sommer et al. (2014)
6. Fertilizer use		

Adapted from Nhamo, N., Rodenburg, J., Zenna, N., Makombe, G., Luzi-Kihupi, A., 2014b. Narrowing the rice yield gap in East and Southern Africa: using and adapting existing technologies. Agricultural Systems 131, 45–55.

- Weed management and labor availability: Labor demand increases for most smallholder farms under CA and hence its management is critical. Similarly knowledge and experience of the use of herbicides will be a good investment, which could reduce labor for weeding in CA plots.
- Mechanization/equipment: CA driven by exclusively manual labor has its limitations and hence the need for mechanization of CA fields. In this regard, both the knowledge of use and the capacity to procure equipment are important.

CA is one of the many ways of managing resources on the farm to reduce erosion, build resilient soil systems, and improve productivity. It has components that directly contribute to both mitigation and adaptation to climate change; however, targeting has been poor in the past. The emphasis of taking it to small farmers without reasonable resource to invest on their farms seems to be misplaced. Consideration of the within-farm factors and the environments outside the farm is an important step that will determine the result of popularization and scaling out efforts. Fig. 7.2 summarizes some of the issues that need further analysis to describe carefully which farmer typology to target for CA adoption.

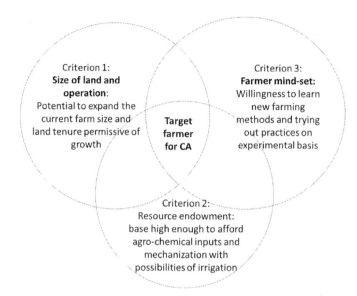

Figure 7.2 An illustration of some of the factors that describe a farmer who can potentially rip benefits by applying conservation agriculture practices.

In addition, farmers are a heterogeneous group and categories range from those who essentially work as laborers on farms of other farmers to those who are wealthy and invest their income from agriculture in business. Furthermore, there are those who get income from outside agriculture and are often considered as part-time farmers who need more knowledge on improved crop production technologies. Therefore, CA cannot be viewed as a solution for all smallholder farmers in southern Africa.

7.4 ECOLOGICAL INDICATORS OF SUSTAINABILITY

Sustainable agricultural technologies deliver ecosystem services that are public good for consumption by communities. CA contributes to increased ecosystem services, especially by the effect of the management practices on soil quality. Ecosystem services are defined here as components of nature that are directly enjoyed, consumed, or used to yield human well-being, for example, purification of air and water, mitigation of droughts and floods, generation and preservation of soils and renewal of their fertility, detoxification and decomposition of wastes, pollination of crops and natural vegetation, dispersal of seeds, cycling and movement of nutrients, and control of the vast majority of potential agricultural pests (Boyd and Banzhaf, 2007). Other slightly different definitions include (Fisher et al., 2009) the conditions and processes through which natural ecosystems, and the species that make them up, sustain and fulfill human life (Daily, 1997); the benefits human populations derive, directly or indirectly, from ecosystem functions (Costanza et al., 1997); and the benefits people obtain from ecosystems (Nelson et al., 2009). CA can be applied on large scale and ecosystem services will be supported. However, its current application on small fragmented plots may not yield much with regard to ecosystem services.

In CA systems, there is sustainable management of the soil and this improves its general fertility. Umar et al. (2013) showed that CA fields that had incorporated *F. albida* trees recorded an increase in organic matter content and available nutrients. This has an indirect benefit of reduced costs for farmers. Additionally, farmers can utilize natural ways of replenishing the soil through the crop rotations embedded in CA. By doing so, they have increased the capacity to achieve household food and nutritional security.

7.5 RESEARCH GAPS

Research for devolvement demands that in evaluating agricultural systems, our view has to be multidisciplinary, driven by the need for sustainability and profitability to reduce the effects of disaster emanating from poverty, land degradation, and climatic extremes. Despite the development of CA in many areas, several of the following questions remain unanswered and these are under agronomic, institutional, and socioeconomic spheres:

1. *Permanent soil cover*: Use of maize stover as the major source of soil cover conflicts with competing uses of the resource. The strategy would be to find unpalatable cover crops to livestock for use where communal grazing systems are practiced. Fencing of plots, controlled grazing, and stall feeding could be alternative options that are to be considered. Unless alternative cover crops are found that are not preferred by animals or improved management practices are developed, CA will have little chance of success in mixed crop-livestock systems.

2. *Spatial variability within fields*: Soil heterogeneity is high in smallholder farming systems. Taking into account this fertility gradient, farmers through extension and cooperatives can invest in new and affordable ways of mobile soil testing in conjunction with the Soil Science laboratories at the Universities and Agriculture Research Institutions in southern Africa to maximize the benefits of CA through appropriate

management by prioritizing the soil needs. This will in turn benefit more farmers, as it can result in a shift in government cost in input support system by supplying inputs as required by specific soils, foregoing the blanket recommendation.

3. *Soil biology and hidden benefits of CA*: Sustainable utilization of soil as a resource requires a close attention to the development of soil health. Two important measurements that are conspicuously missing in the discussion of benefit of CA are (1) the characterization of the soil biological groups involved and (2) the dynamics of the underground food web on CA plots.

4. *Institutionalizing CA*: Recent research suggests that increased contact with extension staff and involvement of women farmers in CA increases the chances of CA adoption (Mlenga and Maseko, 2015). There is need for research to define which institutional arrangement and gender-friendly equipment should be targeted for increased adoption.

5. *Policies supporting CA*: A whole range of policies and policy frameworks that support the practices and reward those contributing to environmental services by the use of CA are required to popularize the benefits of CA at local, national, and regional scales. In light of climate extremes, there is a need to create an environment where incentives exist for smallholder farmers to use CA to improve production, adapting to and mitigating the impacts of climate change. Current adaptation practices are limited in scope and extent for them to bring benefits and reduce negative externalities of food production (Nhamo et al., 2014a,b). Areas of improvement and policy interventions include land tenure, access to and support for implements, markets, and market linkages and pricing.

7.6 CONCLUSIONS

CA has the potential to increase yields and environmental services on smallholder farmers' fields. However, there is evidence that the performance of CA depends on supportive agronomy such as fertilizer application, improved management of farm operations including timely application of operations, and increased knowledge of the use of agrochemicals, e.g., herbicides and pesticides. These cannot be applied by individual resource-constrained farmers.

The contribution of CA to SOM, infiltration, and soil fauna activities was recognized by a number of studies. However, the benefits are limited by application on small plots. Climate-smart, labor-reducing strategies are required, as CA on smallholder farms is currently labor intensive. More data is required on the greenhouse gas emissions from CA plots. Smallholder farmers require knowledge, resources for equipment procurement, large landholding, and agronomy knowledge to effectively move CA forward.

REFERENCES

Anderson, J.A., D'Souza, S., 2014. From adoption claims to understanding farmers and contexts: a literature review of Conservation Agriculture (CA) adoption among smallholder farmers in southern Africa. Agriculture, Ecosystems & Environment 187, 116–132.

Arslan, A., McCarthy, N., Lipper, L., Asfaw, S., Cattaneo, A., 2014. Adoption and intensity of adoption of conservation farming practices in Zambia. Agriculture, Ecosystems & Environment 187, 72–86.

Bationo, A., Mokwunye, A.U., 1991. Alleviating soil fertility constraints to increased crop production in West Africa: The experience in the Sahel. In: Mokwunye, A.U. (Ed.), Alleviating soil fertility constraints to increased crop production in West Africa. Springer, Netherlands, pp. 195–215.

Baudron, F., Mwanza, H.M., Triomphe, B., Bwalya, M., 2007. Conservation Agriculture in Zambia: A Case Study of Southern Province. African Conservation Tillage Network (ACT), Kenya.

Baudron, F., Tittonell, P., Corbeels, M., Letourmy, P., Giller, K.E., 2012. Comparative performance of conservation agriculture and current smallholder farming practices in semi-arid Zimbabwe. Field Crops Research 132, 117–128.

Blarel, B., Hazell, P., Place, F., Quiggin, J., 1992. The economics of farm fragmentation: evidence from Ghana and Rwanda. The World Bank Economic Review 6 (2), 233–254.

Boyd, J., Banzhaf, S., 2007. What are ecosystem services? The need for standardized environmental accounting units. Ecological Economics 63, 616–626.

CFU, 2006. Reversing Food Insecurity and Environmental Degradation in Zambia through Conservation Agriculture. Conservation Farming Unit, Lusaka.

Classen, A.T., Sundqvist, M.K., Henning, J.A., Newman, G.S., Moore, J.A.M., Cregger, M.A., Moorhead, L.C., Patterson, C.M., 2015. Direct and indirect effects of climate change on soil microbial and soil microbial-plant interactions: what lies ahead? Ecosphere. 6 (8), 130. http://dx.doi.org/10.1890/ES15-00217.1.

Costanza, R., d'Arge, R., deGroot, R., Faber, S., Grasso, M., Hannon, B., Limburg, K., Naeem, S., O'neill, R.V., Paruelo, J., Raskin, R.G., 1997. The value of the world's ecosystem services and natural capital. Nature 387, 253–260.

Daily, G.C., 1997. Nature's Services. Island Press, Washington DC.

Derpsch, R., 2008. No-tillage and conservation agriculture: a progress report. In: Goddard, T., Zoebisch, M.A., Gan, Y.T., Ellis, W., Watson, A., Sombatpanit, S. (Eds.), No-Till Farming Systems. World Association of Soil and Water Conservation, Bangkok, pp. 7–39 (special publication no. 3).

Devkota, K.P., Lamers, J.P.A., Manschadi, A.M., Devkota, M., McDonald, A.J., Vlek, P.L.G., 2015a. Comparative advantages of conservation agriculture based rice–wheat rotation systems under water and salt dynamics typical for the irrigated arid drylands in Central Asia. European Journal of Agronomy 62, 98–109.

Devkota, M., Devkota, K.P., Gupta, R.K., Sayre, K.D., Martius, C., Lamers, J.P., 2015b. Conservation agriculture farming practices for optimizing water and fertilizer use efficiency in Central Asia. In: Drechsel, P., Heffer, P., Magen, H., Mikkelsen, R., Wichelns, D. (Eds.), Managing Water and Fertilizer for Sustainable Agricultural Intensification. International Fertilizer Industry Association (IFA), International Water Management Institute (IWMI), International Plant Nutrition Institute (IPNI), and International Potash Institute (IPI), Paris.

Dobermann, A., 2004. A critical assessment of the system of rice intensification (SRI). Agricultural Systems 79 (3), 261–281.

Elwell, H.A., Stocking, M.A., 1988. Loss of soil nutrients by sheet erosion is a major hidden farming cost. Zimbabwe Science News 22 (7/8), 79–82.

FAO, 2005. The Importance of Soil Organic Matter; Key to Drought-Resistant Soil and Sustained Food Production. FAO Soils Bulletin 80. Rome, Italy.

FAO, 2010. FAO Statistical Database – Agriculture. Rome. http://www.faostat.fao.org.

Fisher, B., Turner, R.K., Morling, P., 2009. Defining and classifying ecosystem services for decision making. Ecological Economics 68 (3), 643–653.

Fofana, B., Wopereis, M., Zougmore, R., Breman, H., Mando, A., 2003. Integrated soil fertility management, an effective water conservation technology for sustainability dryland agriculture in sub-Saharan Africa. In: Beukes, D., de Villiers, M., Mkhize, S., Sally, H., van Rensburg, L. (Eds.), Proceedings of the Symposium and Workshop on Water Conservation Technologies for Sustainable Dryland Agriculture in Sub-Saharan Africa (WCT). Bloemfontein.

Giller, K.E., Andersson, J.A., Corbeels, M., Kirkegaard, J., Mortensen, D., Erenstein, O., Vanlauwe, B., 2015. Beyond conservation agriculture. Frontiers in Plant Science 6, 870.

Giller, K.E., Winter, E., Corbeels, M., Titonell, P., 2009. Conservation agriculture and smallholder farming in Africa: The heretics' view. Field Crops Research 114, 23–34.

Guto, S.N., Pypers, P., Vanlauwe, B., de Ridder, N., Giller, K.E., 2011. Tillage and vegetative barrier effects on soil conservation and short-term economic benefits in the Central Kenya highlands. Field Crops Research 122 (2), 85–94.

Hagmann, J., 1994. Lysimeter measurements of nutrient losses from a sandy soil under conventional-till and ridge-till in semi-arid Zimbabwe. In: Merckx, R., Mulongoy, J. (Eds.), The Dynamics of Soil Organic Matter, Proceedings of the 13th International Conference on Soil Tillage for Crop Production and Protection of the Environment. Aalborg.

Hagmann, J., 1996. Mechanical soil conservation with contour ridges: Cure for, or cause of, rill erosion? Land degradation & development 7, 145–160.

Hobbs, P.R., 2007. Conservation agriculture: what is it and why is it important for future sustainable food production? Journal of Agricultural Science 145, 127–137.

Holland, J.M., 2004. The environmental consequences of adopting conservation tillage in Europe: reviewing the evidence. Agriculture, Ecosystems & Environment 103 (1), 1–25.

Ibragimov, N., Evett, S., Esenbekov, Y., Khasanova, F., Karabaev, I., Mirzaev, L., Lamers, J., 2011. Permanent beds versus conventional tillage in irrigated arid Central Asia. Agronomy Journal 103 (4), 1002–1011.

IIR and ACT, 2005. Conservation agriculture: A manual for farmers and extension workers in Africa. International Institute for Rural Reconstruction Nairobi and African Conservation Tillage Network, Harare, p. 250.

IPCC, 2007. Climate Change: Impacts, Adaptation and Vulnerability. In: Parry, M.L., Canziani, O.F., Palutikof, J.P., van der Linden, P.J., Hanson, C.E. (Eds.), Contribution of Working Group II to the Fourth Assessment Report of the Intergovernmental Panel on Climate Change. Cambridge University Press, Cambridge.

Jayne, T.S., Rashid, S., 2013. Input subsidy programs in sub-Saharan Africa: a synthesis of recent evidence. Agricultural Economics 44 (6), 547–562.

Karungi, J., Adipala, E., Kyamanywa, S., Ogenga-Latigo, M.W., Oyobo, N., Jackai, L.E.N., 2000. Pest management in cowpea. Part 2. Integrating planting time, plant density and insecticide application for management of cowpea field insect pests in eastern Uganda. Crop Protection 19 (4), 237–245.

Kladviko, E., 2001. Tillage systems and soil ecology. Soil and Tillage Research 61, 61–76.

Lal, R., 2015. Sequestering carbon and increasing productivity by conservation agriculture. Journal of Soil and Water Conservation 70 (3), 55A–62A.

Lunduka, R., Ricker-Gilbert, J., Fisher, M., 2013. What are the farm-level impacts of Malawi's farm input subsidy program? A critical review. Agricultural Economics 44 (6), 563–579.

Mlenga, D.H., Maseko, S., 2015. Factors influencing adoption of conservation agriculture: a case for increasing resilience to climate change and variability in Swaziland. Journal of Environment and Earth Science 5, 16–25.

Motsi, K.E., Chuma, E., Mukamuri, B.B., 2004. Rainwater harvesting for sustainable agriculture in communal lands of Zimbabwe. Physics and Chemistry of the Earth, Parts A/B/C 29 (15), 1069–1073.

Mugabe, F., 2004. Evaluation of the benefits of infiltration pits on soil moisture in semi-arid Zimbabwe. Journal of Agronomy 3 (3), 188–190.

Mupangwa, W., Love, D., Twomlow, S., 2006. Soil–water conservation and rainwater harvesting strategies in the semi-arid Mzingwane Catchment, Limpopo Basin, Zimbabwe. Physics and Chemistry of the Earth, Parts A/B/C 31 (15), 893–900.

Mutekwa, V., Kusangaya, S., 2007. Contribution of rainwater harvesting technologies to rural livelihoods in Zimbabwe: the case of Ngundu ward in Chivi District. Water SA 32 (3), 437–444.

Nelson, E., Mendoza, G., Regetz, J., Polasky, S., Tallis, H., Cameron, D., Chan, K., Daily, G.C., Goldstein, J., Kareiva, P.M., Lonsdorf, E., 2009. Modeling multiple ecosystem services, biodiversity conservation, commodity production, and tradeoffs at landscape scales. Frontiers in Ecology and the Environment 7 (1), 4–11.

Ngwira, A., Johnsen, F.H., Aune, J.B., Mekuria, M., Thierfelder, C., 2014. Adoption and extent of conservation agriculture practices among smallholder farmers in Malawi. Journal of Soil and Water Conservation 69 (2), 107–119.

Nhamo, N., 2007. The contribution of different fauna communities to improved soil health: a case of Zimbabwean soils under conservation agriculture PhD thesis. Rheinischen Friedrich-Wilhelms-Universitat, Bonn.

Nhamo, N., Makoka, D., Fritz, O.T., 2014a. Adaptation strategies to climate extremes among smallholder farmers: a case of cropping practices in the Volta Region of Ghana. British Journal of Applied Science & Technology 4 (1), 198.

Nhamo, N., Rodenburg, J., Zenna, N., Makombe, G., Luzi-Kihupi, A., 2014b. Narrowing the rice yield gap in East and Southern Africa: using and adapting existing technologies. Agricultural Systems 131, 45–55.

Nyamangara, J., Chikowo, R., Rusinamhodzi, L., Mazvimavi, K., 2013a. Conservation Agriculture in Southern Africa. Conservation Agriculture: Global Prospects and Challenges, vol. 14, p. 339.

Nyamangara, J., Masvaya, E.N., Tirivavi, R., Nyengerai, K., 2013b. Effect of hand-hoe based conservation agriculture on soil fertility and maize yield in selected smallholder areas in Zimbabwe. Soil and Tillage Research 126, 19–25.

Oicha, T., Cornelis, W.M., Verplancke, H., Nyssen, J., Govaerts, B., Behailu, M., Haile, M., Deckers, J., 2010. Short-term effects of conservation agriculture on Vertisols under tef (Eragrostis tef (Zucc.) Trotter) in the northern Ethiopian highlands. Soil and Tillage Research 106 (2), 294–302.

Pulatov, A., Egamberdiev, O., Karimov, A., Tursunov, M., Kienzler, S., Sayre, K., Tursunov, L., Lamers, J.P., Martius, C., 2012. Introducing conservation agriculture on irrigated meadow alluvial soils (Arenosols) in Khorezm, Uzbekistan. In: Cotton, Water, Salts and Soums. Springer, Netherlands, pp. 195–217.

Scopel, E., Triomphe, B., Affholder, F., Da Silva, F.A.M., Corbeels, M., Xavier, J.H.V., Lahmar, R., Recous, S., Bernoux, M., Blanchart, E., de Carvalho Mendes, I., 2013. Conservation agriculture cropping systems in temperate and tropical conditions, performances and impacts. A review. Agronomy for Sustainable Development 33 (1), 113–130.

Showers, K.B., 2002. Water scarcity and Urban Africa: an overview of urban–rural water linkages. World Development 30 (4), 621–648.

Sommer, R., Thierfelder, C., Tittonell, P., Hove, L., Mureithi, J., Mkomwa, S., 2014. Fertilizer use should not be a fourth principle to define conservation agriculture: response to the opinion paper of Vanlauwe et al. (2014) "A fourth principle is required to define conservation agriculture in sub-Saharan Africa: the appropriate use of fertilizer to enhance crop productivity". Field Crops Research 169, 145–148.

Thierfelder, C., Cheesman, S., Rusinamhodzi, L., 2013. Benefits and challenges of crop rotations in maize-based conservation agriculture (CA) cropping systems of southern Africa. International Journal of Agricultural Sustainability 11 (2), 108–124.

Tursunov, M., 2009. Potential of Conservation Agriculture for Irrigated Cotton and Winter Wheat Production in Khorezm, Aral Sea Basin (Ph.D. dissertation). Universitäts-und Landesbibliothek, Bonn.

Twomlow, S.J., 2000. The influence of tillage on semi-arid soil–water regimes in Zimbabwe. Geoderma 95 (1), 33–51.

Umar, B.B., Aune, J.B., Johnsen, F.H., Lungu, O.I., 2011. Options for improving smallholder conservation agriculture in Zambia. Journal of Agricultural Science 3 (3), 50.

Umar, B.B., Aunes, J.B., Lungu, O.I., 2013. Effects of Faidherbia albida on the fertility of soil in smallholder conservation agriculture systems in eastern and southern Zambia. African Journal of Agricultural Sciences 8, 173–183.

Vanlauwe, B., Wendt, J., Giller, K.E., Corbeels, M., Gerard, B., Nolte, C., 2014. A fourth principle is required to define conservation agriculture in sub-Saharan Africa: the appropriate use of fertilizer to enhance crop productivity. Field Crops Research 155, 10–13.

Verhulst, N., Govaerts, B., Verachtertb, E., Castellanos-Nararretea, A., Mezzalamaa, M., Walla, P.C., Chocobar, A., Deckers, J., 2010. Conservation Agriculture, Improving Soil Quality for Sustainable Production Systems? (Ph.D. (Unpublished)). Food Security and Soil Quality.

Vogel, H., 1993. Tillage effects on maize yield, rooting depth and soil water content on sandy soils in Zimbabwe. Field Crops Research 33 (4), 367–384.

Zougmore, R., Kambou, F.N., Ouattara, K., Guillobez, S., 2000. Sorghum-cowpea intercropping: an effective technique against runoff and soil erosion in the Sahel (Saria, Burkina Faso). Arid Soil Research and Rehabilitation 14 (4), 329–342.

FURTHER READING

Fisher, B., Costanza, R., Turner, R.K., Morling, P., 2007. Defining and Classifying Ecosystem Services for Decision Making. CSERGE Working Paper EDM, No. 07–04.

Mutema, M., Mafongoya, P.L., Nyagumbo, I., Chikukura, L., 2013. Effects of crop residues and reduced tillage on macrofauna abundance. Journal of Organic Systems 8 (1), 5–16.

Wall, P.C., 2007. Tailoring conservation agriculture to the needs of small farmers in developing countries: an analysis of issues. Journal of Crop Improvement 19 (1–2), 137–155.

CHAPTER 8

The Use of Integrated Research for Development in Promoting Climate Smart Technologies, the Process and Practice

Jim Ellis-Jones[1], Therese Gondwe[2], Terence Chibwe[2], Alexander Phiri[3], Nhamo Nhamo[2]

[1]Agriculture-4-Development, Silsoe, Bedfordshire, United Kingdom; [2]International Institute of Tropical Agriculture (IITA), Southern Africa Research and Administration Hub (SARAH) Campus, Lusaka, Zambia; [3]Lilongwe University of Agriculture & Natural Resources, Lilongwe, Malawi

Contents

8.1 WHAT IS INTEGRATED AGRICULTURE RESEARCH FOR DEVELOPMENT?

Investments in agricultural research have been strong components in strategies to promote sustainable agricultural development in most African countries. However, the context has evolved over time. Initially

Smart Technologies for Sustainable Smallholder Agriculture
ISBN 978-0-12-810521-4
http://dx.doi.org/10.1016/B978-0-12-810521-4.00008-6

165

agricultural research focused on strengthening research supply, then shifted to improving links between research and extension. However, these links remained linear with research generating new technologies for extension, who were then expected to transfer these to farmers. The focus has now changed, as it has become apparent that the supply and demand for research were more complex than the linear approaches implied and that the involvement of multiple stakeholders could speed the use of research. The concept of "Integrated Agriculture Research for Development" (IAR4D) is based on the use of an innovation systems approach for implementing agricultural research. This represents an important shift from the traditional ways of conducting research with a clear focus on "research for development." This embraces interactions between stakeholders to encourage both the development and use of research, benefiting a wide range of actors (World Bank, 2007). It brings together multiple actors often along a value chain to address challenges and identify opportunities for generating innovation. The approach creates a network of partners who are able to consider the technical, economic, social, institutional, and policy constraints in an environment that facilitates research and learning, not only generating new knowledge, products, or technologies, but also ensures the use of these research products.

The use of IAR4D based on an innovation systems approach has increased rapidly, with donors, international and regional organizations, national governments, and nongovernmental organizations (NGOs) increasingly seeking to promote stakeholder partnerships from both public and private sectors. The strength of IAR4D lies with partner involvement and participation in building partnerships for finding solutions to challenges and opportunities often centered on commodity value chains, with practical solutions to problems being developed as part of the research process. This engages research users in a process of action research and learning by doing. This contrast to traditional research is linked to implementation with an explicit agenda for change, often regarded as learning cycles based on participatory research and extension approaches (PREA).

The Forum for Agriculture Research in Africa (FARA), the International Livestock Research Institute (ILRI), and the International Institute of Tropical Agriculture (IITA) among others have been key players in developing and promoting IAR4D, using an innovation systems approach based on PREA, bringing partners together within Innovation Platforms (IPs) (Adekunle et al., 2013; Hall, 2006). IITA played an important role in FARA's sub-Saharan Africa Challenge Program with pilot learning sites in Kano, Katsina, and Maradi, Lake Kivu and Zimbabwe, Malawi, and Mozambique (FARA, 2009) and more recently in southern Africa in Malawi, Mozambique, Swaziland, and Zambia on a project "Making Innovations work for smallholder farmers affected by HIV and AIDS (IITA, 2014). In addition, the UK Government, Department for International Development (DFID) "Research-into-Use" (DFID-RIU) program, covering a number of African and Asian countries, was based on an innovation systems approach (DFID-RIU, 2013).

An IP has been described as a space for learning and change, comprising groups of individuals often representing different organizations with different backgrounds and interests, such as farmers, traders, food processors, researchers, extension agents both government and NGO. Members come together to diagnose problems, identify opportunities, and find ways to achieve their goals. They may design and implement activities as a group or coordinate activities undertaken by individual members of the community.

8.1.1 Guiding Principles

IAR4D and innovation system approaches have four defining principles (Hawkins et al., 2009), which integrate:

- The perspectives, knowledge, and actions of different stakeholders around a common theme.
- The learning that stakeholders achieve through working together.
- Analysis, action, and change across environmental, social, and economic dimensions of development.
- Analysis, action, and change at different levels of spatial, economic, and social organization.

Putting these principles into effect requires joint activities, knowledge sharing, joint analysis action, and change. This necessitates individual, organizational, and institutional capacities to ensure that these activities take place. This requires different stakeholders, individuals, and organizations to come together on a level playing field. IAR4D should not be viewed solely as an approach or a framework but rather a basis for ensuring the quality of the processes. It is concerned not only with technology or policy outputs, but also with improving capacity and behavioral processes that ensure that research is put into use. This requires favorable organizational and institutional environments, which may require suitable governance structures, leadership, and management, resources, procedure, and culture to ensure that IAR4D becomes part of mainstream research and development practice.

8.1.2 Roles and Interventions

Interventions to support the innovation process vary depending on the purpose and are influenced by both the context and the capacity of different stakeholders. Fig. 8.1 demonstrates a typical three-phased process from initial engagement with stakeholders, thorough planning, implementing, learning, and assessing until a final phase of ensuring a continuing and sustainable and dynamic innovation system (Devaux, 2005).

At each phase in the innovation process, the role of the participants is likely to change. In the case of local participants this changes from one of interest to active collaboration and finally ownership and leadership. At the same time the role of research and development organizations needs to change from one of initial leadership to facilitation of the process and finally to one of providing backstopping, when and as required. The role of the private sector is likely to mirror that of local participants in changing from interest to one of active collaboration and finally to that of farmer support

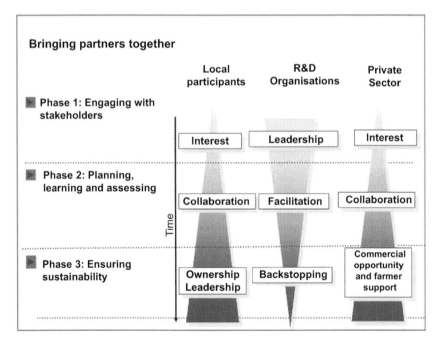

Figure 8.1 Conceptual framework for innovation platforms approach, establishment, and functioning. *Modified from Devaux, A., 2005. Strengthening Technical Innovation Systems in Potato-Based Agriculture in Bolivia. R 8182. Final Technical Report for Department for International Development (DFID)'s Crop Protection Programme. International Potato Center (CIP), Lima, Peru.*

Box 8.1 Typical Interventions at Each Phase of the Innovation Process

Phase 1: Engagement With Stakeholders
- Building and supporting partnerships
- Creating common vision, trust and awareness raising
- Building capacity to understand problems and identify opportunity
- Developing attitudes, practices, and incentives

Phase 2: Planning, Learning, and Assessing
- Assessing input and output markets, value chain analysis
- Developing action plans for systems improvement, value addition, and market opportunity
- Agreeing partner roles
- Innovation research and development
- Learning, assessing performance, and capacity development
- Enhancing collaboration across actors and sectors

Phase 3: Ensuring Sustainability
- Setting in place new innovations (products, technologies, management practices, institutions, marketing, and policies)
- Ensuring ownership by local participants
- Maintaining agility and ability to identify new opportunities
- Providing backstopping as required

and commercial opportunity. Interventions can occur in any of the three phases, key ones being shown in Box. 8.1.

One of the most important activities that IPs should undertake is to build the capacity of its members to innovate. This is a crucial function as innovation capacity for IPs is key to their aims. Key elements of innovation capacity include self-organization, learning new skills, changing behavior, and mindsets, valuing others' roles, having a holistic view, being able to adapt to changing situations, creating new ideas, recognizing opportunities, being proactive, using local ideas, and looking to the future.

An IP is not necessarily meant to last forever. Once underlying problems have been solved, the IP should not be kept alive artificially. However, the innovation capacity that develops the thought process can and should live on, underpinning its importance as a key output of the IP. Enhanced innovation capacity is one of the most sustainable outcomes that IPs should strive for.

Innovation will not happen by research alone. It requires new knowledge and new ideas from different individuals to be combined to arrive at new solutions. Communication helps IP members to identify common objectives. IPs should help to manage information and ensure institutional memory. IPs ensure that all members' voices are heard giving them ownership, clarifying agendas, and visions for change that help in bringing members together.

8.2 COMPONENTS OF IAR4D ACTION RESEARCH USING PARTICIPATORY RESEARCH AND EXTENSION APPROACHES

IPs generally follow several consecutive steps (ILRI, 2013), namely:

- **Initiation**. Any stakeholder group can initiate an IP, but it is usually a research or development organization, a government agency or an NGO that does so. This organization identifies the broad focus, identifies stakeholders, and brings them together convening the first few meetings. It also identifies someone to facilitate the establishment and early operations of the IP.
- **Deciding on focus**. IP members discuss the focus area, challenges, and opportunities. Members gather information from various sources including research, current practice, local knowledge, and relevant policy guidelines. Community engagement and situation analysis are often undertaken to facilitate some of the information.
- **Identifying options**. IP members decide what they want to do to resolve the challenges or take advantage of opportunities that they have identified. This might be new drought-resistant crops or varieties, better use of inputs, improving marketing or processing, or pressing for policy change.
- **Testing and modifying solutions**. Solutions must be tested and adapted to make sure they work. The IP coordinates these experiments and monitors whether they are successful. The results of the experimentation feed into the innovations used by the IP members.
- **Developing capacity**. In most cases developing the capacity of different actors is essential for innovation to succeed. Farmers may need training, cooperatives or associations may need help with organization, leadership, and management, support may be needed with producing seed or marketing. The IP needs to identify these gaps and find ways to develop the capacity required.
- **Implementing and scaling up**. If the innovation is successful, the IP works with member groups to encourage adoption. This may mean

documenting and communicating the innovation, arranging training, and study visits and persuading other groups to adopt in a process of peer-to-peer extension. Farmer Field Schools and ongoing action research can play an important role in this process.

- **Learning**. Learning what has succeeded and what may have failed are an important role of IPs. Often this can be undertaken during field days, end-of-season evaluations, and process reviews. The information should be fed back to IP members, so that they can identify any further changes that may need to be made.

IPs represent a systematic means of encouraging change through joint action. Although they are institutional structures they need to be flexible and able to change in response to changing situations (ILRI, 2013), in particular:

- **Changing focus**. As problems are solved and new issues arise, the activities and focus of the IP may change. Such an adaptation is an important step in learning as a group.
- **Changing membership**. IP membership may change as needs arise. The IP may invite new members or bring in outsiders on a short-term basis to provide information or advice.
- **Changing responsibilities**. Management of the IP should shift over time from the initiating organization to one or more members. For example, a farmer organization may take over responsibility for organizing the IP from a government agency or research body.
- **Temporary or permanent**. IPs may be temporary, existing only as long as necessary to resolve the problem or may continue as new challenges and opportunities arise. If so, it is necessary to find ways to fund the IP's activities, after initial funding ends.
- **Linking to other organizations**. IP members provide links to the organizations or groups they represent with each member communicating the IP activities and suggestions back to his/her own organization.
- **Linking IPs**. Some problems cannot be addressed at one level. It may be useful to form IPs at several levels to address problems at each one. IPs at different levels should be linked to share ideas and information. For instance, National or District-level IPs are more likely to operate at a strategic level and local or community level. Different groups of farmers at local level can be linked through innovation clusters.

PREA mirrors and reinforces the action research embodied in IAR4D. It comprises four interlinked phases representing a learning cycle, often covering a full cropping season. The phases include:

- Engaging with stakeholders, identifying constraints, and raising awareness
- Identifying opportunities and action planning
- Implementing through experimentation or demonstration
- Learning and reviewing

Facilitation and capacity building are key to ensuring sustainability and ownership of both the process and results by IP members.

8.2.1 Partnerships and Innovation Platforms for Crop Value Chains

IAR4D and innovation system approaches are often based on improvements in commodity value chains in which knowledge, purchased, and farm-provided inputs are used in natural resource-based production systems, marketed, and processed for sale and consumed. A value chain can be regarded as a full range of activities required to bring a product or service from conception, through the different phases of production (involving a combination of physical transformation and the input of various producer services), delivery to final consumers, and final disposal after use (Kaplinsky and Morris, 2001). A typical value chain comprises an entire system of production, processing, and marketing from inception to finished product (Table 8.1). Value chain actors include researchers, farmers, traders, processors, wholesalers, retailers, and consumers linked together by flows of products, finance, information, and services. Value chain supporters such as government regulators, financial institutions, research and extension, transporters provide services to the chain enabling it to function.

This approach requires identification of the actors involved at all stages along the chain, followed by a systematic analysis to identify the challenges and opportunities for ensuring fair reward for all and in particular producers, who are often the main target beneficiary. Innovation can be shaped in different ways, depending on the initial context and whether the key actors are public or private and whether they operate at International, Regional, National, District, or Local Government or Community levels. IPs operating at higher levels of governance and management hierarchies are likely to be concerned with policy and strategy issues and might include chief executives or directors of stakeholder organizations to agree strategies to promote innovation along a value chain. They can also facilitate the activities of IPs

Table 8.1 A typical crop and seed value chain indicating functions, process, and actors

Functions	Research	Regulation	Seed testing and evaluation	Seed production, promotion and sales	Farm production	Household use and local sales	Adding value and marketing
Process	On station and on-farm trials undertaken in conjunction with seed testing and evaluation	Variety approvals, training, and certification	PB and PVS PREA[a]	Production of "certified" and "quality declared" seed	Farmer adoption of new varieties and management practices	Health and nutrition training New recipe training	Introduction of small and medium scale mechanization and processing equipment
Actors	International and National Research Institutes and Departments	National Seed Institutes	Farmers Govt. extension NGOs	Community-based seed producers *Seed companies* Cooperatives Agro-dealers	Farmers	Govt. extension NGOs Nutrition promoters Households	Engineers and fabricators Farmers and farmer groups Traders and agro-processors Consumers

NGO, nongovernmental organizations.

[a]Participatory Breeding (PB) and Participatory Varietal Selection (PVS) linked with Participatory Research and Extension Approaches (PREA).

operating at local or implementation levels. IPs established at these levels source membership from the same stakeholders, but targeting farmers, farmer groups, and frontline staff who have the mandate of their organizations.

8.3 IMPLEMENTING IAR4D: THE CASE OF ESTABLISHING CASSAVA INNOVATION PLATFORMS IN ZAMBIA AND MALAWI

8.3.1 Experiences From Zambia

In Zambia, where IPs are a relatively new concept, five district level cassava value chain IPs were initiated in 2013 by the IITA through a project, Support to Agricultural Research for Development of Strategic Crops in Africa (SARD-SC). This was implemented in five districts across three Provinces of Zambia. The process involved identification of key stakeholders along the cassava value chain, including farmers, traders, transporters, agro-dealers, agricultural extension staff, processors, bankers, researchers, and traditional leaders.

IITA facilitated workshops in each district aimed at raising awareness about SARD-SC, creating a common vision and building capacity of stakeholders on the use of PREA to enhance inter-stakeholder collaboration. During initial meetings, facilitation aimed at promoting interaction among those stakeholders having different but complementary roles along the value chain. Participants undertook a participatory diagnosis of challenges and identification of opportunities along the cassava value chain.

Challenges included: low staffing of qualified breeders and agronomists, poor links between research and extension, limited access by farmers to improved cassava planting material, limited processing knowledge and skills, limited availability of cassava processing equipment, limited financing opportunities, poor market linkages, and a lack of policy support for the cassava sub-sector. Opportunities included: income generation from multiplication of improved cassava varieties, value addition through cassava processing, use of existing appropriate technologies, and employment creation through enhanced activity in the cassava sub-sector.

At the end of each workshop representatives from each of the five district level value chains, either volunteered or were nominated with each establishing a local IP committee to undertake the following:
- Co-ordinate research and development activities
- Identify challenges and opportunities for agricultural innovation and development
- Encourage interaction between public and private sectors, NGOs, and Community Based Organizations

- Act as a strategic entry point for agreed interventions
- Hold regular meetings, at least one meeting per quarter (making a total of 4 meetings in a year at the minimum)
- Plan, implement, assess progress during field days and review activities at the end of the cassava cropping season.

IP members were nominated on a voluntary basis with a contact person nominated to serve as a liaison officer with IITA, to assist in dissemination of information to other IP members and arrange regular IP meetings. Meeting information was communicated through either the IP secretary or the IP chairperson, to send out phone messages to all IP members to inform them of meeting dates, venues, and agendas.

After successful launch of the five districts IPs, each was requested to elect a substantive executive committee to develop guidelines for their operations. This included capacity building for the development of their activity work plans based on SARD-SC project activities, encouragement to work with other stakeholders, including the Zambia Agricultural Research Institute and the Seed Control and Certification Institute in the implementation of their activities.

8.3.2 Lessons From Innovation Platforms

At the time of writing, the IPs were in the second phase of development as shown in Fig. 8.1: planning, learning, and assessing. Although all five IPs remain dependent on the guidance and support of IITA, two performed better than others, meeting regularly and engaging other value chain stakeholders. These included small and medium enterprise processors, who had been trained in making confectionary products from cassava flour and exhibiting their products at district agricultural shows. Farmers from three IPs have been involved in the provision of improved disease-free clean cassava cuttings to meet requests from farmers. Ministry of Agriculture extension staff played a key role in identifying farmers with improved cassava cuttings. Local transporters also members of the IPs provided transport for delivering cassava cuttings to farmers across the three districts. Other IPs have been slower in operationalizing their action plans.

Challenges affecting the IPs include:

- Most IP meetings recorded low attendance following the launch.
- Although IP meeting frequency varies, agreed meeting periods were often not achieved.
- Although IPs remain informal, there have been requests by some for formal identification and registration to enable them to operate bank accounts and participate in other activities along the value chain as formal entities.

Because the IPs were not always fulfilling all the roles expected of them, refresher training was provided at which IP chairpersons and secretaries participated in a 2-day training aimed at improving the understanding and role and operations of their IPs.

8.3.3 Experiences From Malawi

The concept of IPs was introduced in Malawi by DFID's M-RIU Program. M-RIU first established commodity-based innovation platforms along aquaculture and legumes value chains as a means of bringing stakeholders together to agree on commodity improvement strategies, hence embracing innovation systems thinking (Moyo, 2013) and addressing key challenges in their sectors. Lessons learnt from these IPs contributed to the establishment of other IPs along other value chains, including the Cassava Commodity Innovation Platform.

Cassava is a staple food crop for some 30% of the 17 million people in Malawi, more especially for those living in the central and northern regions along the lake shore, where the soils are generally poor (Moyo et al., 1998). A recent review (Phiri, 2016) revealed that although cassava is an important crop among many Malawian households, the crop has received little political attention. The review noted that although the Government had since independence made a deliberate effort to make maize "a political" crop in all aspects, very little support had been provided to other important crops including cassava. It is perceived by many as a "poor person's" crop often relied on for food security, if maize fails. This is despite its potential and success as a cash crop in many countries in West Africa.

The increasing effects of climate change, resulting in erratic rainfall patterns and increasing frequency of droughts and dry spells, have necessitated diversification from maize. Additionally, the declining performance of the tobacco market, the main foreign exchange earner of the economy provides opportunity for other crops including cassava to receive greater attention. One major area of cassava that has enjoyed substantial institutional support is research and extension, thereby introducing more players into the subsector. Cassava has through a number of stakeholders enjoyed increased publicity through the media, field days, and demonstrations. This has ultimately called for establishment of a coordination mechanism to encourage knowledge sharing, cross-fertilization of ideas, and minimize duplication of effort to avoid waste of valuable and scarce human, material, and financial resources. Because the primary targets of most activities are farmers, there is a need for all the actors in the subsector to convey common messages. Additionally, by operating through a well-coordinated structure, good practices can be shared with all stakeholders to the benefit of the value chain.

This coordination mechanism was the Cassava Platform, established at the national level. This has now evolved into the Roots and Tuber Crops Innovations Development Trust (RTCIDT), established in 2010 and providing an IP for stakeholders. RTCIDT was established by relevant national stakeholders in the sector, who identified the need to strengthen root and tuber crop production and productivity, value addition and marketing to elevate importance of these crops for food security as well as potential for investment at the national level.

Thus the main objectives of the RTCIDT were as follows:

- To facilitate linkages and coordination of stakeholders and activities for root and tuber crops development including identifying technological gaps, building stakeholder capacity promoting research needs, and other improved interventions to address the identified gaps
- To enhance increased investment for market development of the sector to address food security, nutrition and industrial product issues (starch, glue, animal feed, etc.), and advantages with regard to import substitution
- To advocate at the political level for promotion of improving business operational environment
- To establish an advisory, information, and knowledge-sharing service to support interventions of relevant players and users of root and tuber crops
- To establish the Trust (IP) as a permanent nonprofit dialog and advocacy forum

The Trust (or IP) grew not only to become a platform for sharing ideas and good practices, but also slowly growing into a strong policy instrument for the promotion of cassava in the country. It is believed that as more investors are attracted into the subsector through activities of the Trust, the image of cassava as a poor man's crop will slowly fades.

8.4 BENEFITS AND CHALLENGES OF IAR4D AND IPs

Although the potential benefits of IAR4D and IPs have been well documented in this chapter, there are a number of challenges that need to be addressed either before IP establishment or certainly during its deliberations (Table 8.2). Ensuring buy-in from members requires commitment of time and other resources, which should ideally be obtained before IP

Table 8.2 Benefits and challenges associated with IAR4D and IPs

Benefits	Challenges
• Creates a forum for dialog and understanding • Facilitates communication on an equal basis • Enables identification of constrains and opportunities • Leads to better informed decisions • Creates motivation and feeling of ownership • Contributes to capacity development • Makes innovative action orientated research possible • Enhances adoption and impact	• Progress and success requires buy-in from all members • Can be difficult and costly to establish • Requires long-term perspective engaging actors and developing relationships over time • Danger in IPs being regarded as forums for transfer and dissemination of technologies • Tangible outputs are necessary to sustain membership and commitment to the innovation process • Can be difficult to monitor and evaluate in a systematic way

IAR4D, Integrated Agriculture Research for Development; *IPs*, Innovation Platforms.

establishment through prior discussion with potential stakeholders. Cost sharing is essential for sustainability, even when these may initially be borne by one partner. Commitment is often developed as relationships develop and results are achieved. Tangible results will help in sustaining membership and ensuring sustainability of the innovation process. There is a danger that IPs become part of a bureaucratic process where individual stakeholders push their own agendas and the IP merely becomes a forum for technology promotion in a linear approach rather than through a coalition of innovative stakeholders. In these situations IA4RD and IPs are unlikely to be successful. Monitoring and evaluating their achievements is, therefore, important, so that corrective action to ensure innovative thinking takes place.

IPs are intended to generate multiple interactions among the players. The more successful the discussions and business transactions that are generated during IP interactions, the more successful the platform can become. Fig. 8.2 shows some examples of the connections platform members can make as they participate in the IP activities.

8.5 LESSONS LEARNED FOR SCALING UP

It is increasingly recognized that IAR4D and innovation system approaches can play a major role in introducing new ways of working. Interventions to encourage innovation depend on the initial context and how this changes over time. Interventions should not focus only in developing research

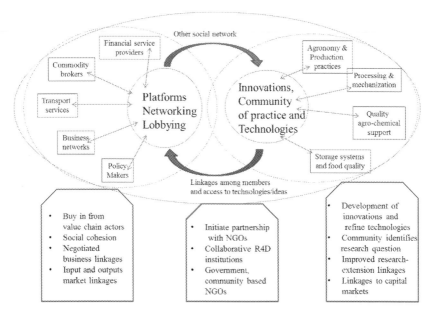

Figure 8.2 An illustration of the multiple interactions involved when innovation platforms are fully functional as players seek to use the platform to develop a community of practice.

capacity, but rather analytical and negotiation skills among the different players. Development of interactions should be encouraged among public, private, NGO, and community-based organizations. Although research is likely to be an important component, it is often not the central one and in the early stages, interventions to build capacity, access, and use existing knowledge and foster learning are likely to be important. Research questions should come through feedback loops created during the implementation of IPs activities. Clearly ready and timely access to inputs including finance is crucial. This needs to be based on effective and competitive marketing and at the same time addressing social and environmental concerns.

Key elements that support IAR4D include the following:

1. *Improving infrastructure*: This should include roads, communication, and power, which provides the basis for ensuring inputs can be made available at affordable prices and outputs delivered to market. This is often a precursor to adding value along market chains. Success has been linked to communication and accessibility of the IP committee members.

2. *Building and supporting partnerships*: Creating a network of stakeholder groups drawn from both public and private sectors is a prerequisite. Such groups need to have the capacity and be able and willing to interact and

work together in an environment that encourages cooperation, build trust, and establish a common vision for the future. This will require awareness raising, the development of trust, a willingness to work together, and creation of a shared vision for the future.

Facilitating or brokering alliances is critical and incurs an indispensable and unavoidable cost, which is often overlooked. Such alliances also require "champions," either individuals or institutions, who understand the often complex institutional and regulatory structures that underpin, encourage, and support the building of networks.

3. *Innovation platforms*: An innovation platform composed of partner organizations represents a strong approach to empowering participating stakeholders, building capacities, and identifying opportunities able to analyze, alleviate constraints, and add value to a value or systems chain.

4. *Strengthening farmer organizations*: Ensuring the establishment and participation of effective and representative farmer organizations able and willing to communicate with members. In most cases this will require support and capacity development to create strong farmer organizations able to speak with an informed and unified voice and able to engage with other stakeholders at all levels on an equal basis. Creation of a local knowledge bank should be a planned as a result of the effective interactions.

5. *Involving the private sector and ensuring use of market driven approaches*: A well-organized private agri-business sector needs to be involved not only in the supply of inputs and purchasing outputs but also in developing market opportunities, capacity building, and engaging with the public and NGO sectors.

   This is likely to require enabling public policies and regulations including deregulation of markets while ensuring competition and compliance with minimum standards have laid a solid foundation.

6. *Improving access to information, knowledge and training*: New knowledge from research is only one component required to encourage innovation in agriculture. Improving access to information can create an effective demand for research products. For instance, use of local radio programs will compliment training, knowledge sharing, and other learning events. If this involves suppliers, technical experts, farmers, government, and NGOs in the radio programs, this will help to build partnerships and networks.

7. *Scaling up and adding value to country agricultural strategies*: The complexity of scaling up successful pilot initiatives can be supported by national stakeholder platforms, linked, and interacting with local or district

platform initiatives. Target platforms could be those with similar socio-economic environments.

8. *Sustainability*: This requires capacity strengthening throughout the process to ensure local people and organizations assume ownership and leadership. This should be continuous and not undertaken as a one-off activity, requiring long-term funding commitment. Ultimately sustainability is built on ownership by local participants with effective back up from research and development organizations in both the private and public sectors. Sustainability should also be viewed in light of the development of successful enterprises by IP members and exit of the facilitating organizations.

Key issues that need to be addressed in scaling up include better targeting of poor and vulnerable groups especially women; finding sustainable methods of promoting development of rural financial services; and the conscious inclusion of capacity building of IAR4D beneficiaries in efficient management of productive assets (Adekunle et al., 2013).

8.6 CONCLUSION

1. IAR4D is an evolving concept, where IPs have numerous advantages once effectively established.
2. Community engagement and situation analysis should shape the direction and influence the effectiveness of IP coordinating committees in articulating development challenges and opportunities.
3. Effective communication among stakeholders is key for IP development.
4. Ongoing evaluation is required from IAR4D case studies where members reap benefits from participation in IP activities, notably those relating to increased productivity and improved livelihoods.

REFERENCES

Adekunle, A.A., Ayanwale, A.B., Fatunbi, A.O., Agumya, A., Kwesiga, F., Jones, M.P., 2013. Maximizing Impact from Agricultural Research: Potential of the IAR4D Concept. Forum for Agricultural Research in Africa (FARA), Accra.

Devaux, A., 2005. Strengthening Technical Innovation Systems in Potato-Based Agriculture in Bolivia. R 8182. Final Technical Report for Department for International Development (DFID)'s Crop Protection Programme. International Potato Center (CIP), Lima, Peru.

DFID-RIU, 2013. The UK's Department for International Development, Research into Use Programme. http://www.researchintouse.com.

FARA (Forum for Agricultural Research in Africa), 2009. SSA-CP (Sub-Saharan Africa Challenge Programme): Research Plan and Programme for Impact Assessment. Forum for Agricultural Research in Africa (FARA), Accra.

Hall, A.J., 2006. Public-private sector partnerships in a system of agricultural innovation: concepts and challenges. International Journal of Technology Management and Sustainable Development 5 (1), 3–20.

Hawkins, R., Heemskerk, W., Booth, R., Daane, J., Maatman, A., Adekunle, A.A., 2009. Integrated Agricultural Research for Development (IAR4D). A Concept Paper for the Forum for Agricultural Research in Africa Sub-Saharan Africa Challenge Programme. FARA, Accra. 92 pp.

IITA, 2014. Making Innovations Work for Smallholder Farmers Affected by HIV and AIDS. Final Report. International Institute of Tropical Agriculture.

ILRI, 2013. What Are Innovation Platforms? Innovation Platforms Briefs. International Livestock Research Institute.

Kaplinsky, R., Morris, M., 2001. A Handbook for Value Chain Research, vol. 113. IDRC, Ottawa.

Moyo, N., 2013. Institutional arrangements and innovation brokerage in the Malawi aquaculture and legumes innovation platforms. In: Getting New Aquaculture and Legumes Technologies into Use: A Paper Presented at an International Conference on Innovation Systems for Resilient Livelihoods: Connecting Theory to Practice (26–28 August 2013), Johannesburg, South Africa.

Moyo, C.C., Benesi, I.R.M., Sandifolo, V.S., 1998. Current Status of Cassava and Sweet Potato Production and Utilization in Malawi. Ministry of Agriculture and Irrigation.

Phiri, M.A.R., 2016. Review of Policies and Institutional Framework that Affect Fresh and Processed Cassava Trade in Malawi. FAO. Unpublished Report.

World Bank, 2007. Enhancing Agricultural Innovation. How to Go Beyond Strengthening of Research Systems. The World Bank, Washington, DC.

CHAPTER 9

Taking to Scale Adaptable Climate Smart Technologies

Jim Ellis-Jones[1], Alexander Phiri[2], Terence Chibwe[3], Therese Gondwe[3], Nhamo Nhamo[3]

[1]Agriculture-4-Development, Silsoe, Bedfordshire, United Kingdom; [2]Lilongwe University of Agriculture & Natural Resources, Lilongwe, Malawi; [3]International Institute of Tropical Agriculture (IITA), Southern Africa Research and Administration Hub (SARAH) Campus, Lusaka, Zambia

Contents

9.1 WHAT DOES TAKING TO SCALE MEAN?

Taking to scale or scaling-up and scaling-out are terminologies often used synonymously to refer to a wide adoption of new technologies and/or innovations. It can also be regarded as getting more benefits to more people more quickly. Menter et al. (2004) separated scaling-up typologies into (1) quantitative scaling-up—increase in the number of people involved through replication of activities, interventions, and experiences; (2) functional scaling-up—expansion of type of activities; (3) political scaling-up—move beyond service delivery and toward structural or institutional change, and (4) organizational scaling-up—organizations improve efficiency and effectiveness to allow for growth and sustainability of interventions. These four can also be summarized into horizontal and vertical scaling-up. Furthermore, scaling-up, which is often incorporated during the planning phase, involves building institutional capacity in the community for promoting and sustaining the innovation and adoption process, information and learning, building linkages, dialog with policy, and sustaining the process (Franzel et al., 2001). The concept of going to scale is illustrated in two figures (Figs. 9.1 and 9.2).

Smart Technologies for Sustainable Smallholder Agriculture
ISBN 978-0-12-810521-4
http://dx.doi.org/10.1016/B978-0-12-810521-4.00009-8

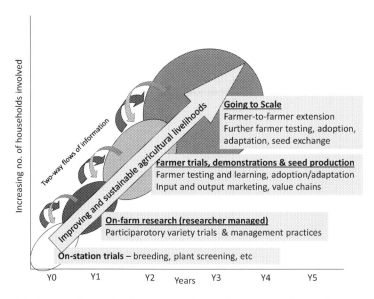

Figure 9.1 A research and development scaling pathway or continuum for smallholder agriculture technologies.

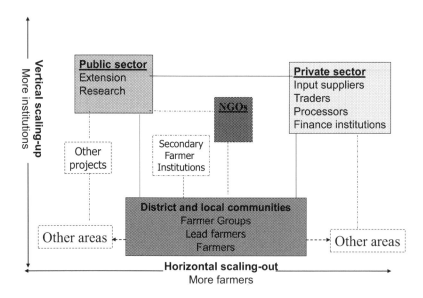

Figure 9.2 Scaling-up and scaling-out illustration showing both the vertical and horizontal components.

Fig. 9.1 illustrates a typical research and development scaling pathway often associated with Integrated Agricultural Research for Development (IAR4D) seeking to improve rural livelihoods through the introduction, in this case, of a new crop variety. This moves from on-station research involving crop breeding and plant screening, progressing to on-farm participatory crop variety trials and evaluations often managed by researchers that may also include natural resource management practice evaluations. This then progresses to less formal but equally important farmer testing, demonstration, and learning opportunities, before finally going to scale, where farmer-to-farmer extension becomes important in ensuring that other farmers can test the new ideas and, if necessary, modify or adapt these to their own circumstances.

During this process communication and feedback between stakeholders (researchers, extension workers, and farmers and the private sector) remain important in developing and finetuning technologies. This helps in ensuring that farmer preference, resource availability, consumer, and market demand are all considered. Innovation Platforms (IPs) can play a key role in building stakeholder partnerships, increasing capacity, and taking technologies and innovations to scale. A recent research study (FAO, 2016) shows that

- Partnerships with public actors and civil society have an important role in creating linkages between farmers and markets and will encourage increased private sector participation.
- Social and institutional innovations are as essential as technological innovations in transitions to sustainable food systems. This includes multiactor IPs and community-supported agriculture.
- Autonomy, reciprocity, and recognition of the diverse types of knowledge that are fostered through institutional innovations all creating incentives for the adoption of sustainable practices.

Fig. 9.2 differentiates between scaling-up and scaling-out, the latter being shown on the horizontal axis with more farmers in different communities and areas adopting a new technology.

The vertical axis demonstrates scaling-up, as more institutions, public and private sector, as well as nongovernment organizations (NGOs) and secondary farmer institutions (associations, cooperatives, Community-based Organisations (CBOs), etc.) become sources of information and advocates of the new technology. Developing interinstitutional linkages and building partnerships early in the IAR4D process play an important role in taking technologies to scale. This also depends on information availability and the extension approaches used.

9.2 THE EVOLUTION OF EXTENSION APPROACHES

Until comparatively recently, extension in much of rural Africa consisted of farmers and communities being told what to do. Often, when governments imposed their concepts of agricultural development, this was characterized by issuing recommendations or instructions. The underlying concept was that scientific knowledge was superior to farmers' knowledge and farmers were encouraged to adopt new technologies because they had been developed by scientists. Those who adopted were seen as innovators and those, who did not, were seen as laggards. An extension officer's job was to convince farmers of the development potential of the new technology and encourage adoption.

Criticism of this approach was based on a lack of adoption by farmers. Critics saw it as top-down, context-less, and scientifically based. In reaction, scientists reexamined the technologies to determine where the technology may have gone wrong. Consequently, farming systems research became increasingly important. This focused on understanding and improving existing systems rather than their wholesale replacement (Table 9.1). Technical change was premised on an understanding how farmers perceived and managed their farming systems with extension centered on ensuring farmers had access to inputs. This systems approach resulted in increased emphasis on on-farm trials and technology development under more realistic conditions. Although the approach gave more credence to the farmer, farmers' input was still limited to help identify problems rather than seeking solutions. Although much of the philosophy of this approach remains relevant, in practice, it has been high cost, often with an inability to scale-up recommendations and sometimes resistance by research and extension to be drawn out of traditional modes of operation.

As understanding of a farming systems approach was realized, Training and Visit (T&V) extension systems were being promoted in many countries. These endeavored to ensure that field staff concentrated on extension without the burden of other activities such as administering subsidy schemes or distributing inputs. Nevertheless, T&V still retained a strongly hierarchical structure with field-based extension workers backed by subject matter specialists relying on strong technical messages reminiscent of technology transfer. As a result T&V systems became both costly and consequently often nonfunctional.

Table 9.1 Research and extension trends since the 1960s to date

Period	Explanation for nonadoption	Prescription	Extension activities	Research methods
1950–60	Ignorance	Research	Technology transfer	Commodity research
1970–80	Farm level constraints	Remove constraints	Training and Visit (T&V) supply inputs	Constraints analysis farming systems research
1990–2000	Technology does not fit	Change process provide options	Facilitate farmer participation	Enhancing farmer competence changing professional behavior
2010	Commodity value chains not effective	Integrated agricultural research for development	Multiple stakeholder market-led approaches	Farmer involvement in research design, implementation, and evaluation

Adapted from Chambers, 1993. Challenging the Professions: Frontiers for Rural Development. Intermediate technology Publications, London, UK; Ellis-Jones, J., Schulz, S., Chikoye, D., de Haan, N., Kormawa, P., Adedzwa, D., 2005. Participatory Research and Extension Approaches. A Guide for Researchers and Extension Workers for Involving Farmers in Research and Development. Research Guide 71. International Institute of Tropical Agriculture, Ibadan, Nigeria.

With the demise of T&V, priority shifted to encouraging increased farmer participation in identifying challenges and opportunities, in equal partnerships with researchers and extension workers and increasingly private sector input suppliers, output marketers, and processors. In accepting farmers as participants, it became clear that farmers also experiment and adapt technologies to their own situations. Science was no longer seen as a privileged knowledge set but simply as a source of information. Consequently, emphasis is being given to finding solutions that start with farmers, through discussion between farmers, extension workers, and scientists. The contrast between "Traditional" and "Participatory" approaches is shown in Table 9.2.

Among the major structural challenges in the development of extension and research in southern Africa has been a separation of government

Table 9.2 Comparison of traditional and participatory extension approaches in smallholder agriculture

	Traditional extension approach	Participatory extension approach
Main objective	Technology transfer	Building farmer capacity, empowerment
Needs analysis and prioritization	Outsiders	Farmers, facilitated by outsiders
Role of private sector	Often not considered state provided services	Engagement from the start of the process
Research results	Fixed packages, messages	Options considered
Farmer behavior	Adopt, adapt, or reject	Choose from basket and experiment
Intended outcomes	Widespread adoption of package	Wider choices, enhanced adaptability
Main extension mode	Extension worker to farmer	Farmer to farmer
Role of extension	Teacher, trainer	Facilitator, provider of choice

extension and research departments, leading to a lack of communication between the two. Poor operational funding and sometimes political interference have led to extension personnel often diverting their attention away from extension activities. Consequently their impact has not been consummate with the numbers and their presence in the communities that they serve. A realigned and reframed extension and research workforce could serve the needs of farmers better. Involvement of both research and extension in IPs would improve communication, collaboration, trust, and motivation, reducing competition and conflicts of interest between the two groups. However, reforms in research and extension alone without funding commitments from governments may also not be as effective.

9.3 PARTICIPATORY RESEARCH AND EXTENSION APPROACHES

Participatory research and extension approaches (PREA) (Ellis-Jones et al., 2005; Hagmann et al., 1999) involve farmers in a continuous process from definition of a research and development agenda, conduct of research, evaluation of results, and promotion of findings. This requires facilitation of local communities' own analysis of their farming systems, identification of constraints and challenges, and the search for solutions and new opportunities. This requires the building of strong working partnerships between local communities, extension agents, researchers and the private sector, and encouraging farmer–to–farmer extension of appropriate technologies and new knowledge. The partnerships established during the process can be regarded as IPs, with initially research and development (R&D) agents providing leadership with active participation of local communities and the private sector. In time ownership and leadership should move to local communities with the R&D organizations continuing to provide back-up support services. Ongoing participation by the private sector will largely depend on commercial opportunity. Such partnerships or platforms should survive beyond the life of the project and contribute to the sustainability of project achievements.

PREA mirrors and reinforces the action research embodied in IAR4D (Ellis-Jones and Gondwe, 2014). It is a process for integrating and improving the effectiveness of both research and extension. Its fundamental principles include the acceptance that farmers are both "experimenters" and "practitioners," that Extension Agents become "facilitators" of change, and that local communities and farmers "own" experiments and demonstrations with "indigenous" knowledge and local farming practices being recognized as the starting point for change.

PREA can be viewed as a learning cycle with four key stages usually tied to an agricultural season that moves from initial community and stakeholder engagement to identify challenges and opportunities to one of action planning, implementation followed by assessment, learning, and review. This assists in setting innovations in place and importantly starts the scaling process. Learning cycles can be repeated over a number of seasons as joint learning takes place, knowledge is gained, and sustainability realized.

The four phases in the PREA learning cycle (Fig. 9.3) comprise the following:

- Phase 1: Community engagement and mobilization, facilitating communities to analyze their situation and identify and prioritize challenges and opportunities.
- Phase 2: Community level action planning based on the opportunities identified to overcome the challenges.
- Phase 3: Implementation through trying out new ideas through farmer testing and demonstration.
- Phase 4: Sharing experiences, learning lessons, and reviewing the process allowing modification for the second and subsequent learning cycles.

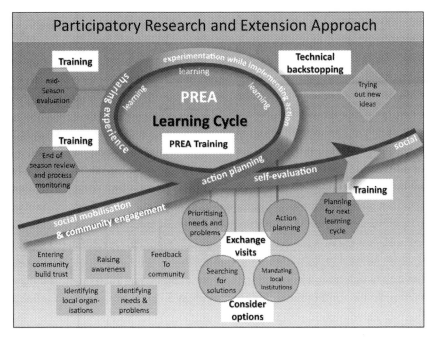

Figure 9.3 Participatory research and extension approach (PREA) learning cycle for agriculture systems. *Modified from Hagmann, J., Chuma, E., Murwira, K., Connelly, M., 1999. Learning Together through Participatory Extension. A Guide to an Approach Developed in Zimbabwe. Department of Agricultural, Technical and Extension Services, Harare, Zimbabwe.*

The complexity of taking to scale successful pilot initiatives can be supported by national stakeholder forum or platforms, linked and interacting with local or district initiatives. These will be important for:

- *Facilitating or brokering alliances*: This can empower stakeholders, build capacity, and identify further opportunities and lead to the establishment of IPs (Adekunle et al., 2013).
- *Improving access to information, knowledge, and training*: This can create an effective demand for research knowledge. For instance, use of well-designed posters and local radio programs will compliment training, knowledge sharing, and other learning events. If this involves suppliers, technical experts, farmers, government, and NGOs in the radio programs, this will help to build partnerships and networks.
- *Strengthening farmer organizations*: This often requires support and capacity development to ensure that farmers can speak with an informed and unified voice and engage with other stakeholders on an equal basis.

- *Involving a well-organized private agri-business sector.* This includes not only in the supply of inputs and purchasing outputs but developing market opportunities, capacity building, and engaging with the public and NGO sectors. This is likely to require enabling public policies and regulations including deregulation of markets while ensuring that competition and compliance with minimum standards have laid a solid foundation.
- *Improving infrastructure*: This is often a precursor to adding value along market chains.

9.4 WORKING WITH LOCAL COMMUNITIES AND THEIR NETWORKS

For development activities to be owned by the community, two key conditions need to be in place: first, real motivation and enthusiasm, and second, effective community-based organizations that can support the process and take it forward. Facilitation and capacity development are important in ensuring sustainability and ownership of both the process and results. Communities and the networks to which people belong play an important role in influencing agricultural practices. It is, therefore, necessary to understand these and where possible incorporate them into a technology development and scaling process. For instance, community-based organizations are ideal for encouraging local involvement and for sharing knowledge allowing communities to use existing networks and to use these to build up new networks. Farmer-to-farmer extension is premised on the belief that for a farmer "*seeing is believing*" and other farmers are the best educators. Through discussions with other farmers and groups they will be stimulated to try out and adopt new technologies. Awareness of farmers' roles as resource managers is increasing and that natural resource management goes beyond merely consulting with farmers to sharing decision making and responsibility for outcomes from management choices and decisions.

Fig. 9.4 illustrates typical linkages between extension agents, researchers, and the private sector with a farmer group or community-based organization. Central to these organizations is a lead or research farmer, selected by the group to test and/or demonstrate the new technology and become a focus for participatory evaluations and training in a farmer field school approach. At the same time he/she should be encouraged to link with other farmers both within and outside the group in a process of farmer-to-farmer

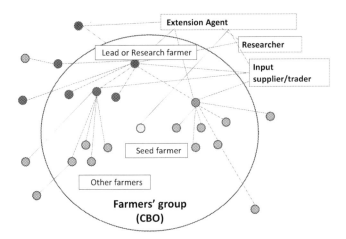

Figure 9.4 Farmer-to-farmer community network and the relationship with research and extension partners for smallholder agriculture.

extension and knowledge sharing. The guiding principles for such activities could include colearning and positive competition leading to refinement and adaptation of technologies.

9.5 LOOKING TO THE FUTURE

Increasingly organizations recognize that agricultural development processes are influenced by the interaction of both environmental and socioeconomic factors, such as trade liberalization, the demands of global markets, urbanization, climate change, agricultural intensification, concentration and integration of value chains, as well as food safety standards and the need to ensure equitable benefits to the actors along value chains (IAASTD, 2009; World Bank, 2012; FAO, 2014; TAP, 2016). There is general agreement that, to meet these challenges, agricultural innovation requires bringing together actors from both the agricultural sector and beyond to share and benefit from their different perspectives and experiences. This requires capacity strengthening both of individuals and organizations, within a policy environment that actively promotes innovation.

Unfortunately interventions are frequently designed and implemented independently, often small in scale and sometimes ending up in taking contradictory positions. At the same time capacity development interventions are often narrow in scope, neglecting institutional capacity and not ensuring learning across systems. They often lack high-level political and operational mechanisms to ensure the comprehensive and sustained effort essential for successful capacity development (TAP, 2016). IAR4D within an innovation system offers opportunity for interaction between policy makers and other stakeholders in ensuring support for capacity development. At the same time coordination and harmonization of approaches would help in promoting improved use of resources of different stakeholders.

Within this framework, an innovation system for agriculture (Fig. 9.5) can be seen as four interlocking components: research and education, bridging institutions, business and enterprise, and the enabling environment. These involve all the key actors (farmers and farmers' organizations, agribusiness, processors, marketers, transporters, input suppliers, policy makers, regulatory agencies, researchers, service providers, extension services, civil society organizations, and others) directly or indirectly in agricultural production, processing, marketing, distribution, and trade.

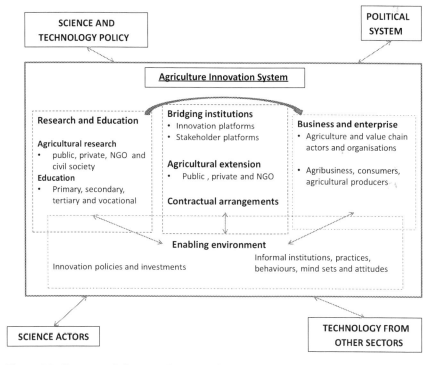

Figure 9.5 Conceptual diagram of an agricultural innovation systems. *Reproduced from Tropical Agriculture Platform, 2016. Common Framework on Capacity Development for Agricultural Innovation Systems: Synthesis Document. CAB International, Wallingford, UK.*

Agricultural innovation also requires space for networking and facilitation of interaction between actors. The processes should aim at building trust and mutual understanding, stimulating learning, and creating the conditions for joint decision making and action that encourages innovation. Strengthening innovation systems goes beyond the exchange and use of knowledge. It needs also to foster entrepreneurship, developing a vision for change, mobilizing resources, building legitimacy, and overcoming resistance to change (Devaux et al., 2011).

Key principles and changes promoted by TAP include using knowledge generation to achieve change; understanding the relationships between the parts of a system including "soft systems analysis," negotiating the meaning of the system and desirable transformations; realizing that participation requires facilitation, engagement for interactive learning between stakeholders, resulting in joint analysis, planning, and action; working with others in ad hoc teams and partnerships; learning how to learn changing from individual learning to social learning and of key importance a shift in the culture of R&D organizations from an exclusive focus on individual merit and competition to promoting collaboration and teamwork within and between organizations.

It is recognized that access to knowledge and information remains the bedrock of an agricultural innovation system. This includes both the assessment of biophysical and market information for the supply of inputs and marketing of produce. Although access to the Internet remains limited, mobile phone networks are increasingly well positioned to deliver information to farmers, enabling them to make better informed decisions and investments that can increase productivity and improve livelihoods. Increased access to mobile phones presents opportunities for farmers to access not only knowledge but also affordable financing.

For instance in Kenya, WeFarm is trying to ensure that every farmer has easy access to information (http://wefarm.org/). This is an SMS-based, peer-to-peer knowledge sharing platform that enables farmers to share and source information. Farmers are able to ask agricultural questions and receive answers directly on their phone. Information generated on WeFarm is also helping companies and organizations working with farmers to tailor their produce and services to suit farmers' needs. Information can also be used to map diseases or outbreaks and help manage those occurrences.

With regards to affordable finance for small farmers, F3 Life is enabling the provision of "climate-smart" or "green" finance by companies, NGOs, and financial institutions to farmers, fishermen, and forest users (http://www.f3-life.com). Under a climate-smart or green financing system, credit scores, interest rates, and credit limits are set with reference to adoption of improved environmental practices by the credit user, as well as conventional credit criteria. The environmental criteria frequently have either climate adaptation and/or mitigation benefits. Presently the system focuses on improved soil conservation and agroforestry measures for increased soil fertility and improved quality and quantity of water in watershed rivers. In the future it is envisaged that systems could underpin a greater variety of improved natural resource management practices and an integrated range of credit instruments designed to build climate-smart sustainable value chains. The F3 Life ecocredit system is underpinned by three tools: first, a system to assess the credit worthiness of small-scale producers, which incorporates environmental measures into a credit-scoring approach. Second, software for disbursement and repayment of loans using mobile money that requires no cash handling by staff with the status of loans being accurately seen on a daily basis. Third, loan tracking can record evidence of adoption of improved agricultural practices or natural resource management, in compliance with the terms of credit. In Kenya, Farmers Life East Africa farmers receive credit on their phones using Kenya's M-Pesa mobile money system. To qualify for loans at reduced interest rates, farmers must plant grass strips and trees across the contour of their land, protecting their farms from soil erosion. Preliminary data indicate that although conventional environmental and natural resource management projects result in less than 15% annual rate of uptake of new practices or technology, F3 Life's approach has to date resulted in a much higher uptake.

GSMA mAgri Program is forging partnerships between mobile operators, technology providers, and agricultural organizations, supporting scalable mobile services that impact farmers and the agricultural industry (http://www.gsma.com/mobilefordevelopment/).

In Zambia, the Farmer's Union has engaged farmers for the development of a platform for sharing information on prices of agricultural commodities. The challenge is now to move beyond information sharing on commodity prices to providing and improving business opportunity for farmers by reducing transaction costs associated with trade in farm produce. The increasing popularity of use of mobile phones and information technologies present opportunities for increasing services for agriculture. Capacity building at both individual and community levels is essential for this.

In Malawi, discussions are underway to strengthen the demand-driven and pluralistic extension delivery system. It is planned that Village Action Plans (VAP) across the country are updated so that farmer demands, institutional and human capacity available are clearly identified. Once VAPs are well-articulated, agricultural technocrats from both government and private sectors can isolate training needs and develop tailor-made training curriculum. The curriculum will then be used to respond to the articulated community needs and to be used by multiple extension agents to train the targeted farmers in the demanded areas for standardization. These targeted farmers then in turn champion farmer-to-farmer extension at village level for greater impact and coverage once they master the technologies learnt at household level and so become lead farmers.

Capacity development of farmers is key in the advancement of demand-driven extension systems (Parkinson, 2009). Where demand-driven extensions systems have been tried the role of the public sector in shaping this knowledge market agriculture to prevent potential market failure and considerable refinement of the definition of demand by the farmers have been identified as areas that requires consideration (Klerkx et al., 2006). Involvement of farmers in IPs increases the chances of farmers understanding the value chains and eventually their capacity to clearly articulate the kind of extension services they require. Chipeta (2006) also highlighted high motivation, reliable, and profitable market opportunities for smallholder farmers as drivers of demand-driven agricultural advisory. Demand-driven service delivery systems can be provided by the private sector or the public sector (Davis, 2008). The main guiding principles for demand driven advisory have been identified as (1) services that shall be driven by users and service providers that shall be accountable to the users, (2) users shall have a free choice of service and "demand" is defined as what farmers ask for, need, and value so much that they are willing to invest their own resources, such as time and money, to receive the services. However, for a demand-driven agricultural extension to be successful there is need for developing enabling policies and public sector commitment to the transition from the current systems (Birner and Anderson, 2007). More research is required on this concept as there are indications that efficient service delivery can be achieved through this approach.

9.6 CONCLUSIONS

Scaling-up successful pilot initiatives require support from national, provincial, and district fora or platforms that can link and interact with local stakeholders using an innovation systems approach. Traditional top-down R&D approaches have no role in the development of future linkages for farmers, value chain players, and IPs. Although interventions depend on the initial context, they should be developed from the beginning in a way that encourages interaction between public, private, NGO, and civil society organizations. This includes:

- Building and supporting partnerships is a prerequisite that requires engagement and collaboration between stakeholders involving awareness raising, the development of trust, a willingness to work together, and creation of a shared vision for the future. IPs are composed of partner organizations that represents a strong approach to empowering participating stakeholders, building capacities, and identifying opportunities able to analyze, alleviate constraints, and add value to a value or systems chain.
- Encouraging communication and trust among stakeholders starting in the field.
- Promoting learning-by-doing to enhance technical and market knowledge and experience.
- Strengthening farmer organizations to be able to speak with an informed and unified voice and to engage with other stakeholders at all levels has a critical role to play. This requires the participation of effective and representative farmer-based CBOs able and willing to communicate with members.
- Building capacity, accessing existing knowledge, and fostering learning are required. Improving access to information can create an effective demand for new knowledge. If this involves suppliers, technical experts, farmers, government, and NGOs, this in itself will help to build partnerships and networks. Looking to the future the use of Internet and mobile phone technologies can accelerate access to both information and financial services.
- A well-organized private agri-business sector needs to be involved not only in the supply of inputs and purchasing outputs but developing market opportunities, capacity building, and engaging with the public and NGO sectors.

Ultimately sustainability will be built on local ownership with effective back-up from R&D organizations in both the private and public sectors. This requires capacity strengthening throughout the process to ensure that local people and organizations assume ownership and leadership. This should be continuous and not undertaken as a one-off activity, requiring long-term funding commitment.

REFERENCES

Adekunle, A.A., Ayanwale, A.B., Fatunbi, A.O., Agumya, A., Kwesiga, F., Jones, M.P., 2013. Maximizing Impact from Agricultural Research: Potential of the IAR4D Concept. Forum for Agricultural Research in Africa (FARA), Accra.

Birner, R., Anderson, J.R., 2007. How to Make Agricultural Extension Demand Driven? The Case of India's Agricultural Extension Policy, vol. 729. International Food Policy Research Institute, Washington DC.

Chambers, 1993. Challenging the Professions: Frontiers for Rural Development. Intermediate technology Publications, London, UK.

Chipeta, S., 2006. Demand Driven Agricultural Advisory Services. Danish Agricultural Advisory Service (DAAS), Neuchâtel Group, Lindau. http://agris.fao.org/agris-search/search.do?recordID=GB2013202579.

Davis, K.E., 2008. Extension in sub-Saharan Africa: overview and assessment of past and current models, and future prospects. International Food Policy Research Institute 15 (3), 15–28.

Devaux, A., Ordinola, M., Horton, D. (Eds.), 2011. Innovation for Development: The Papa Andina Experience. International Potato Center (CIP), Lima, Peru.

Ellis-Jones, J., Gondwe, T., 2014. Linking Innovations Systems with Participatory Research and Extension Approaches: MIRACLE Experiences from Southern Africa. International Institute of Tropical Agriculture, Lusaka, Zambia.

Ellis-Jones, J., Schulz, S., Chikoye, D., de Haan, N., Kormawa, P., Adedzwa, D., 2005. Participatory Research and Extension Approaches. A Guide for Researchers and Extension Workers for Involving Farmers in Research and Development. Research Guide 71. International Institute of Tropical Agriculture, Ibadan, Nigeria.

F3 Life, 2016. Enabling the Provision of Eco-credit by Companies, Financial Institutions and NGOs to Small-scale Farmers, Fishers and Forest Users. http://www.f3-life.com.

FAO, 2014. Innovation in Family Farming. The State of Food and Agriculture. FAO, Rome, Italy.

FAO/INRA, 2016. In: Loconto, A., Poisot, A.S., Santacoloma, P. (Eds.), Innovative Markets for Sustainable Agriculture – How Innovations in Market Institutions Encourage Sustainable Agriculture in Developing Countries Rome, Italy.

Franzel, S., Cooper, P., Denning, G.L., 2001. Scaling up the benefits of agroforestry research: lessons learned and research challenges. Development in Practice 11 (4), 524–534.

Group Special Mobile – GSM, 2016. Fulfilling the Socio-economic Potential of Agriculture Using Mobile Phones. http://www.gsma.com/mobilefordevelopment/programmes/magri.

Hagmann, J., Chuma, E., Murwira, K., Connelly, M., 1999. Learning Together through Participatory Extension. A Guide to an Approach Developed in Zimbabwe. Department of Agricultural, Technical and Extension Services, Harare, Zimbabwe.

IAASTD, 2009. Agriculture at a Crossroads. Global Report. Island Press, Washington DC, USA.

Klerkx, L., de Grip, K., Leeuwis, C., 2006. Hands off but strings attached: the contradictions of policy-induced demand-driven agricultural extension. Agriculture and Human Values 23 (2), 189–204.

Menter, H., Kaaria, S., Johnson, N., Ashby, J., 2004. Scaling up. In: Pachico, D., Fujisaka, S. (Eds.). Pachico, D., Fujisaka, S. (Eds.), Scaling up and Out: Achieving Widespread Impact through Agricultural Research, vol. 3. CIAT, Cali.

Parkinson, S., 2009. When farmers don't want ownership: reflections on demand-driven extension in sub-Saharan Africa. Journal of Agricultural Education and Extension 15 (4), 417–429.

Tropical Agriculture Platform, 2016. Common Framework on Capacity Development for Agricultural Innovation Systems: Synthesis Document. CAB International, Wallingford, UK.

WeFarm, 2016. Connecting Farmers to Vital Information. http://wefarm.org/what-is-wefarm/.

World Bank, 2012. Agricultural Innovation Systems. An Investment Sourcebook. The World Bank, Washington DC, USA.

CHAPTER 10

Food Processing Technologies and Value Addition for Improved Food Safety and Security

Emanuel O. Alamu, Ackson Mooya
International Institute of Tropical Agriculture (IITA), Southern Africa Research and Administration Hub (SARAH) Campus, Lusaka, Zambia

Contents

10.1 INTRODUCTION

Food processing technologies have advanced at unprecedented levels in tandem with the evolving magnitude and complexity of the global food system. Study of every ancient civilization clearly shows that throughout history humans overcame hunger and disease not only by harvesting food from cultivated land but also by processing it using sophisticated methods. The commitment of food science and technology professionals to advancing the science of food ensuring a safe and abundant food supply and contributing to healthier people everywhere is integral to that evolution (Floros et al., 2010). Unfortunately, these advanced food production/processing technologies have also resulted in some negative externalities that have raised concerns on food safety and food security. Food safety and

Smart Technologies for Sustainable Smallholder Agriculture
ISBN 978-0-12-810521-4
http://dx.doi.org/10.1016/B978-0-12-810521-4.00010-4

security for the rural/farming communities in southern Africa is a major concern, as it directly affects both development and production potentials of the region. In view of the increasing population in Africa, access to nutritious food for the poor households is an important factor to consider in achieving the hunger and malnutrition eradication objectives. Both simple and sophisticated food processing technologies are important in ensuring improved food safety and food security at household levels. Application of science and technology within the food system has the needs of society, in both quality and quantity. Today, questions on food systems center largely around safe, tasty, nutritious, abundant, diverse, convenient, and less costly and more readily accessible food than ever before. Scientific and technological advancements must be accelerated and applied in developed and developing nations alike, if we are to feed a growing world population. The aim of this chapter is to highlight some prevailing technologies in the modern food industry and how they have affected the community-based food processing sector. This chapter will also discuss the efficiency and explicabilities of climate-smart technologies and how they offer solutions.

In southern Africa, value addition is increasingly being dominated by commercial farms and large industries that are heavily dependent on heavy-duty machinery that are energy intensive and heavy capital dependent. The energy crises, largely caused by climate change, that southern Africa is currently experiencing are negatively affecting the food industry. The high energy price has led to the high cost of food production, consequently leading to volatile and hyper food prices. For instance, Zambia and Zimbabwe are facing critical energy crisis due to the regulated electricity generation at the Kariba Hydro Power Station. This is due to the water levels that have drastically decreased in Lake Kariba, hence affecting the generation of electricity at both Kariba North and Kariba South hydro power stations. This has resulted in unprecedented load shedding that has raised the cost of production for both the commercial farms and the food processing industries.

Modern processing industries have subjected the citizenry to the food that is not only highly expensive but also highly refined and lower in nutritional values as certain essential nutrients, roughage, and vitamins are lost during the processing chains. This is a result of a segmented community with different tastes, some from the rural–urban migration and increasing incomes. Traditional methods of processing foods are slowly being lost because people are made to believe that processed foods are more superior to homemade foods; local methods are labor intensive and time consuming and the product has a different standard in terms of quality. Concerns on processed foods are linked to use of preservatives and additives to enhance taste and extend the shelf life of the product. Obesity and food-related illnesses have raised alarm to the public over specific ingredients (particularly salt) that may contribute to illnesses or impact childhood development (Story et al., 2008). Climate-smart technologies are proposing food processing technologies that are applicable to rural communities and in the process will alleviate some of our modern day challenges such as global warming, energy crises, food safety, and food security. Climate-smart technologies have the potential to contribute to food safety and food security.

10.2 FOOD PRODUCTION TECHNOLOGIES IN THE MODERN FOOD INDUSTRIES

Energy, derived from electricity and fossil fuels, plays a major role in the production, processing, and distribution of food. This chapter largely focuses on food processing but here we endeavor to shed some light on food production technologies, as they are the precursors to processing technologies. Production technologies, just like processing technologies, have undergone intense industrialization. Chemical fertilizers, agrochemicals, irrigation systems, and more advanced tractors have pushed agriculture into the modern era and have boosted yields dramatically. The chemical fertilizers made it possible to supply crops with extra nutrients and therefore increase yields. The newly developed synthetic herbicides and pesticides controlled weeds, deterred or killed insects, and prevented diseases, which resulted in higher productivity (Tilman, 1999). However, the use of chemical fertilizers, agrochemicals, and fuel/electricity to power farm machinery poses its own negative externalities. Urea/nitrogen fertilizers are prepared using the Haber–Bosch process, which uses large amounts of fossil energy, mainly natural gas, releasing around 465 Tg of carbon dioxide into the atmosphere each year (International Fertilizer Industry Association, 2009).

Over application of N fertilizer lead to losses into water systems (eutrophication from N and P) and the atmosphere contributing to GHGs leading to global warming. On the other hand pesticides/herbicides residue in food result from inappropriate management. These chemicals pause risk to quality food systems.

10.3 MODERN FOOD INDUSTRIES ARE DEPENDENT ON ENERGY AND THEIR CONTRIBUTION TO CLIMATE CHANGE

The gap between energy needs and access to energy is large, and demand will certainly increase as countries develop. Increasing energy access is essential if the poverty reduction targets set out in the Sustainable Development Goals (SDGs) are to be met. Modern processing industries need energy to transport raw materials and to run the giant machines by way of grinding, mixing, heating, and cooling. Energy is also needed for lighting and heating and for transporting finished products. The forms of energy may vary but the fact is that energy is required and energy in general is expensive. Today there is a correlation between food and energy/oil prices. The prices of both food and energy have risen and fallen more or less in uniform in the recent years. The agrifood sector contributes over 20% of the total greenhouse gas (GHG) emissions, most of which originates from methane and nitrous oxide. There are 2989 billion metric tons of carbon dioxide equivalent (CO_{2e}) in the Earth's atmosphere. About 689 billion metric tons, or 22% of this amount, have been added by human activity. Heavy dependence on fossil fuel has not only affected food prices but also resulted in the pollution of the environment. Industrialization, as can be seen in Fig. 10.1, has contributed to climate change and global warming. Since the 1920s, the global surface temperature has been increasing steadily. The best estimate is that the global average surface temperature has increased by $0.6 \pm 0.2°C$ (IPCC, 2001).

The graph shows that from 1880s to 1920s the variations in temperature were not significant in the Earth's surface temperatures. However, after 1920, the period that coincides with the industrial revolution, the increase in surface temperature was significant. The green revolution is famous because of the advent of the use of chemical fertilizers and agrochemicals. Thus there is correlation between global warming and green revolution. It is therefore plausible to deduce that some agricultural practices have greatly contributed to global warming by polluting the air through emissions coming from agricultural machinery and through the utilization of nitrogen fertilizers. N_2O, a by-product of fixed nitrogen applications in agriculture, is a GHG with a 100-year average global warming potential (GWP) that is 296 times larger than an equal mass of CO_2 (Crutzen et al., 2008). The green revolution resulted in greater productivity and hence called for devising mechanisms of food processing at the industrial level. Our processed food has to be transported down the supply chain from growers to processors to grocery stores and restaurants, and, ultimately, to our plates; energy, fuel, and other resources are used at every step along the way to grow, transport, prepare, pack, cook, and serve the food we eat. This energy and fuel use produces carbon emissions and contributes to climate change, which is the environmental challenge that the United Nations Secretary General Ban Ki-moon calls "the defining issue of our time."

What does climate-smart technology offer as a solution to the challenges related to energy and pollution in the food production/processing technologies? There is an urgent need for the major redesigning of both the food and energy systems. Communities, not multinationals, should become the managers of the global food system to ensure that there is reduction in the dependence on fossil energy inputs while reducing GHG emissions from land-use activities. Achieving this goal will require increasing local food self-sufficiency and promoting less fuel- and petrochemical-intensive

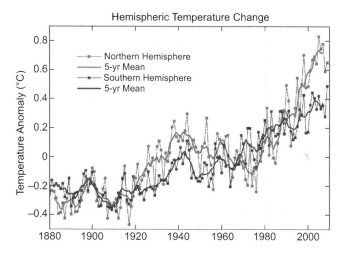

Figure 10.1 Variations of the Earth's surface temperature between 1880 and 2000 for both the Northern and Southern Hemispheres. The Earth's surface temperature is shown year by year [*red bars* (Light gray in print version)] and approximately decade by decade [*blue line* (Dark gray in print version), a filtered annual curve suppressing fluctuations below near decadal timescales]. Over both the past 140 and 1000 years, the best estimate is that the global average surface temperature has increased by 0.6 ± 0.2°C. *From IPCC (Intergovernmental Panel on Climate Change), 2001. Third Assessment Report—Climate Change.* http://www.grida.no/publications/other/ipcc_tar/. http://data.giss.nasa.gov/gistemp/graphs_v3/Fig.A3.gif.

methods of production. Some of the points climate-smart solutions have discovered are listed in the following:

- Waste is a large emission source for food processing. Some of the processing technologies use fossil fuels and so will cause air pollution. However, in food processing, often the majority of the waste is organic, and businesses diverte organics waste to landfill for composting.
- Electricity, although being a relatively small emission source, often corresponds to one of the highest operating costs and offers low-hanging-fruit opportunities for cost savings. However, by the efficient use of electricity, the economies of scale and scope can be achieved, thus lowering the operating cost of using electricity. For instance, in the production of maize meal, food processing industries can endeavor to use machines that will produce more quantities of maize meal per unit time and thus the economies of scale will be achieved. Processing industries can also endeavor to use multipurpose machines that will process more than one product per unit time. For instance, a processing industry can use a machine that will produce both minced meat and sausages at the same time and thus the economies of scope will be achieved.

10.4 CLIMATE-SMART TECHNOLOGIES AND THE FOOD INDUSTRIES

Modern food industries, starting from the farming sector to the processing sectors, are becoming more and more expensive, hence producers' profit margins have significantly decreased over the past years. These changing economic conditions in the food industry have decreased the economic viability of small, and medium-sized farms, reduced the number of farm-related local business and processing facilities, and made the farming profession seem less attractive to the younger generation. In large part, food production has been removed from our communities, thus diminishing the collective knowledge of our region and agrarian practices. Although the current food system offers consumers "inexpensive food," the amount of pollution resulting from modern agriculture is undesirable.

Food processing has existed since time immemorial. In our African traditional communities, food processing included salting, drying, smoking, and fermenting, among others. There are numerous advantages to food processing. Food processing makes many foods available that otherwise would be off-season. Without food processing, we certainly would not have the large variety of food products we see on supermarket and store shelves. Food processing enables the year-round availability of foods that have limited growing seasons. Frozen and canned fruits, vegetables, and meat products are examples. Processing extends the shelf life of foods. Tinned fish and ultra-high-temperature (UHT)-treated milk are two examples of nutritious foods that are readily available as a result of food processing. Food processing also improves food safety by a variety of methods, for example, heating to sufficiently high temperatures destroys harmful bacteria and prevents the growth of harmful fungi and bacteria and packaging helps prevent product tampering. Convenience is another major benefit of foods that have been processed. Imagine not having frozen food or tinned vegetables for that quick and easy Sunday dinner. Some methods of food processing also add flavor to some types of food; for instance, some people prefer dried food to fresh fish.

If less fossil fuel is used for agrochemicals and to power agricultural machinery, our social structures should change, as more farmers will be required to participate in the community-based food production. But for farmers to succeed, current agricultural policies that favor large-scale food production should change in favor of community-based food production/processing. Such policies should be formulated and put in place both by the national and regional governments in consultation with the farming communities.

10.5 CLIMATE-SMART TECHNOLOGIES AND COMMUNITY-BASED FOOD PROCESSING AND INFRASTRUCTURE DEVELOPMENT

The climate-smart technologies support community-based food processing instead of local food processing to emphasize a regional perspective and connect food production/processing with economic and community development. Community-based food processing is a viable form of import substitution that may engage diverse residents. Our definition of a community-based, sustainable food system is consistent with the American Planning Association's definition: "A food system in which everyone has financial and physical access to culturally appropriate, affordable, nutritious food that was grown and transported without degrading the natural environment, and in which the general population understands nutrition and the food system in general" (Cassidy and Patterson, 2008).

Processed foods are convenient especially in urban areas. The current food system places enormous strains on the environment. A more environmentally sustainable food system would require lesser transportation energy and lower fossil fuel inputs, and in accordance with the previous discussion on community-based food production, it would also be climate appropriate. Food production could reuse vacant land, increase vegetative cover, provide opportunities for soil remediation through crop rotation, and increase urban biodiversity by replacing turf grass with a broader array of plants. An environmentally sustainable food system would also recycle waste and restore nutrients in the soil. The promotion of community-based food production will facilitate the development of infrastructures such as processing plants and outlets in which the finished produce will be sold. The community will participate in the food value chain, starting from the farm to the consumers.

10.6 CLIMATE-SMART TECHNOLOGY'S ENHANCEMENT OF FOOD VALUE CHAIN AND MARKET LINKAGES WITHIN THE RURAL COMMUNITIES

The process of growing, producing, and marketing food accounts for most of the activities in the agriculture sector. It involves multiple steps, engages a diverse set of actors, and depends heavily on the operating environment, which is influenced by climate, governance, and other external factors. The multiple steps resulting in the production of food are called the food value chain. The food value chain starts with agricultural input suppliers, which are companies that supply seeds, equipment, and other agrochemicals. Farmers procure the inputs and combine them with other resources needed for agricultural production: land, water, finance, labor, and knowledge. After harvesting, the current trend in Zambia is that farmers sell their produce to middlemen famously known as briefcase traders. The briefcase traders will eventually sell the produce to the processing industries. After industrial-scale processing, packaging, and distribution, the finished products are delivered to consumers through wholesalers and retail outlets. In many third-world countries, in the food value chain, farmers are the most important actors and yet they are the least paid when compared to the other actors.

Sustainability advocates for food value chains that will empower rural communities through market linkages within the rural communities. Directly or indirectly the rural community is dependent on the food value chain. The complexities of the food value chain, and the interdependence of its different components, present both challenges and an opportunities. At present, food value chains in many poor regions are fragmented and inefficient, making them unprofitable and risky for the farming/rural communities. On-farm or community processing may add value to the chain. Farmers sell the finished product directly to consumers or to intermediaries, who in turn trade with consumers, food processors, or traders. Finished products sold locally or within the communities have a relatively short value chain, whereas farm produce that are harvested, processed, and then exported involve a more complex chain.

10.7 CONCLUSION

Modern food production/processing industries and the society seem to be facing tough trade-offs. Agricultural commercial farms have become incredibly good at producing food, but these increased yields have environmental costs that cannot be ignored, especially if the rates of nitrogen and phosphorus fertilization triple and the amount of land irrigated doubles. The tradition in agriculture has been to maximize production and minimize the cost of food, with little regard to its effects on the environment and the services it provides to society. As the world enters an era in which global food production is supposed to double, it is critical that agricultural practices be modified to minimize environmental impacts, even though many such practices are likely to increase the cost of production.

The prudent path toward reforming food processing technologies must coordinate agricultural policies with appropriate technologies. The technologies that we adopt for the production and processing of food must meet the nutritional needs of humanity in a sustainable way. During the 2009 World Summit on Food Security, it was recognized that by 2050, food production must increase by about 70%, which is 34% higher than the current percentage, to feed the anticipated 9 billion people. The intricacies of the modern food industries have made production-to-consumption food systems more insurmountable, thus rendering food safety and food security difficult to attainable. Continued food scarcity invites chaos, disease, and terrorism. Climate-smart technologies will embrace any technology that will guarantee that food is largely safe, tasty, nutritious, abundant, diverse, convenient, less costly, and more readily accessible at the same time mitigate climate change.

REFERENCES

Cassidy, A., Patterson, B., 2008. The Planner's Role in the Urban Food System. The New Planner. http://www.planning.org/thenewplanner/2008/spr/urbanfoodsystem.

Crutzen, P.J., Mosier, A.R., Smith, K.A., Winiwarter, W., 2008. N_2O release from agro-biofuel production negates global warming reduction by replacing fossil fuels. Atmospheric Chemistry and Physics 8, 389–395.

Floros, J.D., Newsome, R., Fisher, W., Gustavo, V., C´anovas, B., Hongda, C., Dunne, C.P., Bruce, J.B., Hall, R.L., Heldman, D.R., Karwe, M.V., Knabel, S.J., Labuza, T.P., Lund, D.B., McGloughlin, M.N., Robinson, J.L., Sebranek, J.G., Shewfelt, G.L., Tracy, W.F., Weaver, C.M., Ziegler, G.R., 2010. Feeding the world today and tomorrow: the importance of food science and technology. Comprehensive Reviews in Food Science and Food Safety 9, 572–599.

International Fertilizer Industry Association, 2009. Fertilizers, Climate Change and Enhancing Agricultural Productivity Sustainably. International Fertilizer Industry Association, Paris, France.

IPCC (Intergovernmental Panel on Climate Change), 2001. Third Assessment Report—Climate Change. http://www.grida.no/publications/other/ipcc_tar/.

Story, M., Kaphingst, K.M., Robinson-O'Brien, R., Glanz, K., 2008. Creating healthy food and eating environments: policy and environmental approaches. Annual Review of Public Health 29, 253–272.

Tilman, D., 1999. Global environmental impacts of agricultural expansion: the need for sustainable and efficient practices. Proceedings of the National Academy of Sciences of the United States of America 96, 5995–6000.

FURTHER READING

Born, B., Purcell, M., 2006. Avoiding the local trap: sale and food systems in planning research. Journal of Planning Education and Research 26, 195–207.

Bridger, J., Luloff, A.E., 1999. Toward an interactional approach to sustainable community development. Journal of Rural Studies 15 (4), 377–387.

Brown, L.R., 2009. Could world food shortages bring down civilization? Scientific American 300 (5), 50–57.

FAO, 2009. Feeding the world, eradicating hunger. In: World Summit on Food Security. November 16–18, 2009. Food and Agricultural Organization of the United Nations, Rome. WSFS2009/INF/2 ftp://ftp.fao.org/docrep/fao/Meeting/018/k6077e.pdf.

Kang, K., Khan, S., Ma, X., 2009. Climate change impacts on crop yield, crop water productivity and food security—a review. Progress in Natural Science 19, 1665–1674.

Magadza, C.H.D., 2000. Climate change impacts and human settlements in Africa: prospects for adaptation. Environmental Monitoring and Assessment 61, 193–205.

Pingali, P.L., 2012. Green revolution: impacts, limits, and the path ahead. Proceedings of the National Academy of Sciences of the United States of America 109, 12302–12308.

Snyder, C.S., Bruulsema, T.W., Jensen, T.L., Fixen, P.E., 2009. Review of greenhouse gas emissions from crop production systems and fertilizer management effects. Agriculture, Ecosystems and Environment 133, 247–266.

Woods, J., Williams, A., Hughes, J.K., Black, M., Murphy, R., 2010. Energy and the food system. Philosophical Transactions of the Royal Society 365, 2991–3006.

CHAPTER 11

Models Supporting the Engagement of the Youth in Smart Agricultural Enterprises

Nhamo Nhamo, David Chikoye
International Institute of Tropical Agriculture (IITA), Southern Africa Research and Administration Hub (SARAH) Campus, Lusaka, Zambia

Contents

11.1 INTRODUCTION

Youth unemployment is a major developmental challenge that spans across many generations, and solutions are yet to be found especially in countries with high population growth rates in Africa. There are nearly 2 billion young adults between the ages 15 and 24 years worldwide, 85% of whom live in developing countries (Sader, 2004). Large numbers of youths either graduate from formal educational institutions armed with school certificates, college diplomas, and university degrees or dropout of schools or leave employment in southern Africa every year. Young adults face a tough job market where there are very few opportunities for meaningful employment. Furthermore,

Smart Technologies for Sustainable Smallholder Agriculture
ISBN 978-0-12-810521-4
http://dx.doi.org/10.1016/B978-0-12-810521-4.00011-6

youth-friendly employment creation in developing countries is an area least developed since the millennium. Youth participation in gainful economic activities at national level and their subsequent contribution to the gross domestic product (GDP) in southern Africa is decreasing.

There are numerous reasons explaining youth unemployment in Africa: poor development planning, quality of education, and vocational skills training that do not focus on appropriate competencies for the job market. Therefore, youths lack the basic skills needed in a competitive labor market (Page, 2012), and furthermore, youth population's growth rate outpaces the creation of decent work opportunities. Limited experience, no professional networks, lack of capital, little professional knowledge, and no savings exacerbate the situation (ILO, 2016; Vogel, 2015). Most youths drop out of school without completing primary or secondary education, which is the main reason why literacy levels have remained low in the region. For those who go through with education the curriculum in colleges does not meet the demands of the labor market (Page, 2012). This scenario has led to youths taking up menial and low-paying jobs, leading to underemployment. A mismatch characterized by unmet expectation from both the youth and the employers has developed and requires attention. This is against a demographic statistical background showing that southern Africa has a youthful population and a very small proportion of old citizens (average age = 18 years; life expectancy = 36 years).

Young people constitute an important segment of the world's demography and excluding them from economic activities increases poverty rates in both urban and rural communities. Regional unemployment figures suggest that youths are increasingly affected by the slowing down of economic performance of most countries in the region (Table 11.1). Low economic performance has affected both public and private sector employment opportunities and hence has reduced youth participation in the economy. A traditional regional structural problem that relates young people to immaturity and incompetence (Sechele, 2015) needs to be challenged if youth development and economic participation is to take shape in the future.

The definition of "youth" varies across different socioeconomic and political institutions worldwide. For instance, youths have been categorized as individuals belonging to the age group of 18–35 years; however, the following are the age variations from this definition: 18–24 years (ILO, 2016), 16–28 years (UNICEF, 2016), and 18–30 years (NEPAD, 2014). In this chapter a broad definition of the youth was adopted as any individual(s) between the age of 18 and 35 years (e.g., Patel, 2009). In addition to their age, youths in Africa

Table 11.1 Youth unemployment figure (%) for 14 Southern African Development Community (SADC) countries for the period between 1996 and 2015

Period	1996–2000		2001–5		2006–10		2011–15		Country average (%)
Country	Female	Male	Female	Male	Female	Male	Female	Male	
Angola	11.5	10.1	11.4	10.1	11.3	10.1	11.2	9.90	10.70
Botswana	39.3	27.0	40.2	27.0	38.8	27.3	39.7	28.5	33.48
DR Congo	13.7	10.7	13.6	10.7	13.6	10.6	13.3	10.3	12.06
Lesotho	48.9	30.1	45.4	29.6	37.3	25.8	40.9	27.5	35.69
Madagascar	6.60	5.8	6.60	5.70	6.20	5.10	5.8	4.60	5.80
Malawi	14.9	12.7	15.1	12.7	15.0	12.7	14.8	12.6	13.81
Mauritius	27.8	17.2	28.2	20.3	29.7	16.0	27.9	16.0	22.89
Mozambique	39.9	41.5	40.0	41.5	40.0	41.5	39.9	41.6	40.74
Namibia	44.9	35.8	40.0	29.3	45.7	36.4	43.3	34.8	38.78
South Africa	55.3	45.6	57.6	47.6	55.5	47.9	57.3	48.8	51.95
Swaziland	46.2	39.4	45.8	39.3	45.6	39.5	46.1	40.1	42.75
Tanzania	7.70	5.40	7.10	4.80	6.50	4.10	6.70	4.40	5.84
Zambia	24.6	28.5	24.6	28.5	23.4	27.4	23.1	27.0	25.89
Zimbabwe	9.30	7.30	9.30	7.30	9.30	8.60	9.40	9.50	8.75
Regional average	27.90	22.65	27.49	22.46	26.99	22.36	27.10	22.54	24.94
Standard deviation	16.79	13.86	16.55	13.75	16.21	14.18	16.57	14.30	15.04

DR Congo, Democratic Republic of Congo.
Source: The World Bank, 2016. The World Bank Group. http://data.worldbank.org/indicator/SP.DYN.LE00.MA.IN/countries.

and other parts of the world are just beginning their life and hence have not had the time and opportunity to acquire wealth. We define wealth here as any form of fixed capital assets, monetary savings, and professional experience that can be used in place of finance and is of value to the owner. Also most youths do not have social capital (White and Gager, 2007), thereby reducing their linkages to job markets. For the purpose of developing economic models that will feature the role of youths in contributing to economic growth, this chapter takes the definition of all young people who have attained a legal age (18 years and older) to qualify for gainful employment as the "youth."

The unemployed youth comprises all persons of working age who are (1) without work, i.e., not in paid employment or self-employment; (2) currently available for work, i.e., available for paid employment or self-employment; and (3) seeking work, i.e., taken specific steps to seek paid employment or self-employment. The specific steps taken to seek employment may include registration at a public or private employment exchange; applying to employers; checking at worksites, farms, factory gates, market, or other assembly places; placing or answering newspaper advertisements; seeking assistance of friends or relatives; looking for land, building, machinery, or equipment to establish own enterprise; arranging for financial resources; and applying for permits and licenses (ILO, 2011).

The aim of this chapter is to review the initiatives by the countries in southern Africa in youth development and engagement for increased employment opportunities and reduction of youth unemployment. It further builds a case for new youth opportunities in transformed agriculture systems and recommends models for validation. The chapter has been arranged as follows: Section 11.2 describes the magnitude of youth unemployment, Section 11.3 analyses the regional youth programs implemented to date across countries, Section 11.4 outlines the developmental opportunities for youths in agriculture, Section 11.5 discusses the technical options available for youth engagement, Section 11.6 focuses examples of youth entry points into agribusiness, Section 11.7 discusses about models to support youth employment creation in agriculture, and Section 11.8 provides the conclusions.

11.2 THE MAGNITUDE OF YOUTH UNEMPLOYMENT AMONG RURAL AND URBAN YOUTHS

Two out of three inhabitants of sub-Saharan Africa are younger than 25 years. Youths account for 65% of the total employment in agriculture, in which they play a key role in agricultural development. Poverty, poor health, hunger, and lack of education limit the potential of the youth in increasing agricultural productivity and incomes (Scorgie et al., 2012; Viner et al., 2012). About 40% of the unemployed are youths and majority of them (70%) live in rural areas. For those who are employed, low productivity, underemployment, and hence meager earnings characterize their employment profile. During the recent 15th anniversary of the Forum for Agricultural Research in Africa (FARA) held on November 2014, in South Africa, policy makers and researchers in Africa have declared the "rising youth unemployment in the continent a potential time bomb that should be quickly defused using appropriate policies and initiatives before it detonates and wreaks havoc on the African continent." Five-year averages on unemployment in the Southern African Development Community (SADC) countries ranged between 4.1% in Tanzania and 57.6% in South Africa (Table 11.1) and these values tend to increase with population increase.

Youth unemployment breeds poverty in communities in southern Africa. Levinsohn (2007) identified (1) potential GDP losses, (2) foregone future growth as the unemployed gain neither experience nor skills, and (3) contribution to social ills as the three main socioeconomic costs of unemployment. Furthermore, unemployment results in wasted talent, brain drain, reduced lifelong personal earnings, increased risk of suicide, and social unrest (Vogel, 2015). A close link has been established among poverty, disease, and malnutrition, leading to high risks in the youth population. Risks include concentration on unproductive and time wasting behavior. As youths seek jobs away from their families, communities, and sociocultural norms, their involvement in activities wasting human capital increases, e.g., use of drugs (including alcohol) and risking sexual activities for money (Scorgie et al., 2012; Viner et al., 2012). An unstable job market also leads to increased mobile workers, leading to high risk of disease spread, e.g., HIV/AIDS. Similarly, youths from a poor family background face challenges with malnutrition-related illnesses and conflicts with colleagues, leading to nonperformance in the workplace. Even more difficult are situations in which school dropouts are driven by the need for youths to care of the sick and elderly because of the limited infrastructure and number of institutions in most countries.

Agriculture remains a key driving force for economic development in the SADC region in which most inhabitants rely on agriculture directly or indirectly as their main source of livelihood. It remains the primary source of subsistence, employment, and income for 61%, or 196 million, of the region's total population of 322 million people (SADC Yearbook, 2013; World Bank, 2015). Agriculture accounts for almost 8% of the region's GDP. Despite the importance of the sector in SADC's economy, agricultural growth rates have been both low and highly variable across the region, averaging only 2.6% per annum in the past decade. Between 1960 and 2005 the net per capita agricultural production decreased by about 40%. This suggests that agricultural production has not kept pace with the population growth in the region, which averaged at 3.2% in 2015. The decline in per capita agricultural production is attributable to several factors including the rapidly growing population in the face of low agricultural productivity. There is a need to identify optimal policy and investment alternatives that will yield the highest payoffs. Governments in the region recently committed to increase the national budgetary allocation for the agricultural sector to 10%, a proposition that could improve agricultural development.

With much higher priority given to implementation of well-designed public investments in agriculture, continued progress on regulatory and policy reform, and modest overlay of inclusion of youth in agriculture, the sector has the potential to reduce poverty and unemployment. However, extreme reliance on government-funded programs and the limited private sector commitment to long-term youth development goals are two major risk factors that youths, especially from remote rural areas, face in southern Africa.

Civic service and volunteer work have largely been implemented in southern Africa. Although youths can benefit by gaining useful experience and employability, lack of incentives, service opportunities, competing livelihood needs, and poverty often limit the number of youth volunteers. Lough and Sherraden (2012) proposed innovative hybrid ways of building assets and financial capabilities while also generating future livelihoods for youths. Such approaches concede that building assets increase financial stability and hence create a focus into a positive future, a line of argument that supports the positive youth development approach.

11.3 REGIONAL CONTEXT AND THE COMMON PROGRAMS IMPLEMENTED ACROSS COUNTRIES

Youth unemployment can be traced back to the days just after independence of most countries in southern Africa, which is some 50–60 years ago. In the period following independence, enrollment into educational institutions increased and this did not match the economic growth of all countries in southern Africa. Therefore, youth unemployment increased as the number of student graduates ballooned with time (Livingstone, 1985; McGrath et al., 2009). Other programs and institutions were initiated in response to youth unemployment during the same period. Vocational education and training programs with a youth employment and self-employment focus were adopted by most SADC countries in the 1980s through the 1990s. These programs were dominated by the Brigades training that accommodated young people aged 18–25 years (Sechele, 2015). The form and design invariably remained similar across the regions and the following are examples of some of the Brigades training programs in different countries: the Tirelo Setshaba in Botswana, Malawi Young Pioneers, National Youth Service in Zimbabwe, Zambia National Service, Namibia National Youth Service, Lesotho National Youth Volunteer Corps Project, and National Youth Service in South Africa (Akoojee et al., 2005; Johanson and Adams, 2004). The main goal of all these programs was to improve youth employability and entrepreneurship skills and to develop skills for self-employment. However, the vocational and technical training evolved gradually and unsystematically and the goals of reducing youth unemployment were lost along the way (McGrath, 2005, 2012; McGrath et al., 2009). This had a large bearing on the quality of both graduates and the labor force joining the different commercial disciplines and eventually on the overall performance of the economies of the countries. In the absence job creation, there was a need to invest in entrepreneurial skills development to move the region forward.

Curtains (2004) defined an "entrepreneur" as an individual who starts and builds a new business. Studies have shown that traditionally entrepreneurship activities are commonly conducted by three age groups: (1) 25–35 years (20%), (2) 35–44 years (15%), and (3) 18–24 years (13%), and a gap of 18.5% exists between the potential for entrepreneurship (25%) and the actual practice (6.5%) (Curtain, 2004; Sechele, 2015). Countries have accepted and promoted small to medium enterprises (Dabalen, 2000) because of the promising gains in creating employment, high returns to financial capital, and combination of skills in the case of people working in groups. The informal nature and the wide range of magnitude of operations in which entrepreneurs are involved has made it difficult for policy makers to formulate suitable policies. This has resulted in the development of weak policies in southern Africa.

The National Youth Service also referred to as the voluntary service was initiated in southern Africa in the 1960s. Its contribution to lifting the poor-income groups, which include youths, women, old citizens, disabled citizens, children, and displaced people, has been invisible. In the SADC, civic service was applied in a propoor, people-centered, human capacity development, and participatory approach. Civic service is defined here as the organized period of substantial engagement and contribution to the local, national, or world community that is recognized and valued by the society, with minimal monetary cost to the participant (Patel, 2009). The National Youth Service programs in southern Africa lacked the multiple dimension definition of youth development and were criticized for focusing on militarizing youths, entrenching power to individual leadership using youth, dispensing patronage, and using youth against general populace. Furthermore, institutional weaknesses drastically reduced the visibility of the outcomes, as bureaucracy and inefficient administration and management reduced funding because of the structural adjustment programs limiting investment in youths and the lack of strategic planning (Kanji, 1995; Patel, 2009).

11.3.1 Recent Initiatives

A number of new initiatives are being implemented in the SADC countries to broadly address the challenge of youth unemployment. The African Youth Decade 2009–18 Plan of Action provides a framework for achieving the objectives of the African Youth Charter. Theme 8.1 of the African Youth Charter seeks to reduce youth unemployment by 2% annually through a number of initiatives, including establishing a comprehensive capacity development program for youths, aligning education curricula to match industry requirements, multiskilling and reskilling of the youth, establishing incubation and internships for youth graduates, and developing national youth employment plans. Most SADC countries are still to achieve the 2% annual reduction in youth unemployment or have put in place a mechanism for eradication, as required in the African Youth Charter (NEPAD, 2014).

11.3.2 Employment-Related Policies

There is a glaring lack of policies related to youth development and youth employment in southern Africa. However, a number of economic and employment, in general, policies have been developed and these have some remote influence on youth unemployment. Table 11.2 summarizes some of the policies found in southern Africa.

There are missing linkages that are common among all the countries in southern Africa and need to be addressed for youth development to take shape (Table 11.2). Failure by regional players to encompass an inclusive and positive youth development approach, including taking developmental programs to the rural remote areas, provides a main and strongest explanation of the current situation of youths in southern Africa (Dabalen, 2000; Sechele, 2015). The following are the developmental areas that can support youth employment and development and also reduce unemployment of young people in the region.

Table 11.2 Some examples of policies that relate to youth development and creation of employment in southern Africa

Policy	Target sector	Relationship to youth employment	Comments
Arable land development policy	Agriculture	Agriculture opportunities are driven by land tenure systems	Botswana
Black economic empowerment of South Africa	Economy	Promotion of economic development results in increase in employment	South Africa
Comprehensive agriculture policy framework	Agriculture	Highlights policy areas needed for agricultural development but no direct mention of youth development	Zimbabwe
Finance assistance policy	Finance	Policy directly affects initial capital limitations in starting businesses	Botswana
Industrial policy of South Africa	Industry	Indirect as industrialization creates some jobs	South Africa
Technical education, vocational, and entrepreneurship training	Education	To create high-quality, sustainable, demand-driven, and equitable training system	Zambia
National youth policy	Youth	Broadly describes the important areas of focus but no concrete steps and timing for action. It includes the desire to include culture, politics, and identity in youth development programs	All the SADC countries, except Seychelles
National skills development policy framework	Labor	Skilled labor issues are articulated not necessarily for youths alone but for the whole labor force	Zimbabwe
Indigenous economic empowerment policy	Economy	No properly defined targets and not inclusive of youths	Zimbabwe
Agricultural policy	Agriculture		All the SADC countries
Land reform policy	Agriculture		South Africa, Zimbabwe

Rail and road infrastructure supportive of business development: Communication is important for the movement of goods and services. In agriculture, access to rail and road networks has an added value on the timely delivery of quality produce to the market. Except for South Africa and Botswana, the rest of the region still has large areas that are not accessible to conduct business transactions. Road transport is the most widely used form of transport in southern Africa, and basic road networks are seldom in existence, making the easy to do business very difficult. Although programs have been initiated by governments under the broad "food for work" initiative, they were not supported with suitable equipment or technology that could be used for decent road repairs and construction and hence the roads were never improved. Opportunities for building centers and rural processing zones were lost because of poor communication networks, especially rail and roads.

After-school training facilities: Other than the vocational training colleges that have been developed in most countries in the region, there has not been a significant effort to develop facilities for youth who finish school with poor grades or drop out of school for various reasons. In the case of vocational training institutions the entry requirements remained purely academic in the sense that the demands for a school certificate remained in force. Largely this has jeopardized opportunities of employment for those who are not academically gifted. Given the resources that are available in most remote areas, informal training and support programs could have helped in resource conservation and sustainable utilization. An example was the campfire project that resolved the wildlife–agriculture conflicts in the boundary zones in Zimbabwe (African Development Bank, 2015). Major lessons stemmed from the improved resource use, community benefits from natural resources, and increased knowledge of both wildlife and agriculture by the lowly educated communities.

Private sector structures at grassroots: The absence of vibrant private activities in remote areas has reduced employment opportunities of rural youth from remote areas in this sector. Most private sector businesses are based in the large cities and hence the rural–urban migration. In southern Africa the transfer of technologies and development, in general, is left to a few nongovernmental organizations (NGOs), with offices in rural areas a scenario which has slowed down development of opportunities in remote areas and hence youth unemployment. A deliberate drive to increase private sector activities in rural areas can increase exposure of rural dwellers and hence increase changes of development.

Information management data processing organizations: Data on youth development and unemployment figures are difficult to find in southern Africa. This is because besides a few schools that are in rural areas, no form of records and data is collected on a regular basis to inform policies. The situation is worse with programs as there is no database to search for programs that the communities have received in the past in order to learn from them. A concerted effort to keep records will feed positively into policy formulation.

Multiple goal-oriented NGOs: Youth opportunities are sometimes varied across time and space. Youth development requires a multifaceted approach, and yet the organizations found in rural areas follow a very narrow range of operations. Quite often the development organizations with multiple goals can contribute to the general development of youth yet they are absent in most countries in southern Africa.

11.4 IDENTIFYING OPPORTUNITIES AND EMPOWERING YOUTH: RESPONDING TO DRIVERS OF YOUTH UNEMPLOYMENT

Youth redundancy was evidently noticed in the early 1980s across southern Africa. Most countries in southern Africa had gained independence, and during the early years of independence, employment for skilled personnel was never a problem. The problem started when structural adjustment programs dictated that public expenditure had to be drastically reduced to meet set targets (Kanji, 1995). In response to increased number of youths who were unemployed, both private and public sectors in southern Africa developed youth training centers, youth clubs, and youth cooperatives. The approach used in these institutions was very much similar to the national military service program meant for young adults that was designed by most developed countries to equip youth with skill sets useful for survival. Limited attention was paid to commercializing ideas that were being promulgated in these institutions. In a sense, the graduates looked forward to be employed in white-collar jobs by government agents or departments after attending the courses, thus avoiding the entrepreneurship route.

A closer look at the quality of training provided by these programs' design for en masse production of graduates shows that the training institutions were poorly equipped, seem to have limited curriculum, and were not geared to solve the impending youth unemployment problem. Similar curriculum irrelevance observed with school and college programs hampered progress. The majority of the programs were similar to an extension of senior-school-level education in style and approach. The need to shift from training everything could have been noticed in the early years of independence and the youth engagement programs could have been initiated then. But the population lost trust and eventually some were closed or converted to other forms of training (McBride, 2009). An inclusive approach with a renewed emphasis on youth development is required to move the region beyond the current situation of unemployment.

In spite of these developments, agricultural extension also lost financial support from many governments at a point when approaches such as training of trainers (ToT) were becoming popular. There was a need to increase the number of farmers reached and dissemination of technologies that improved yields across the region. This approach was augmented by the NGOs working in remote areas. However, often NGOs had parallel arrangements that preferred training their own field personnel. Reaching millions using improved approaches, e.g., peer-to-peer extension and value chain development, has the potential to spread technologies among youth. In the recent years, farmer-driven extension approaches have been proposed; however, limited capacity of farmers to articulate extension needs, a long tradition of free poor-quality extension service, and questions on its sustainability have weighed down these approaches.

11.5 DEVELOPMENTAL APPROACHES FOR ENGAGING YOUTHS IN SMART AGRICULTURE

The future of smart agriculture production depends on sustainable practices adopted widely by farmers, in particular young upcoming farmers. Youth can actively implement modern, highly sophisticated, and high input smart agricultural technologies. The traditional top-down approach to support youth development has not yielded much, as the efforts were led mainly by government departments and the resources were sought just for short-term projects and were often channeled through the long bureaucratic routes with a lot of leakages. Youths were often viewed as a problem that needs to be fixed rather than partners who need to be engaged in the development of communities and agricultural labor force for the future of regional development. This heightened the negative notion that viewed youth as liabilities to national budgets rather than as important individuals who can contribute to the economies of their countries. By default rather than by design, the focus tended to be on the constraints and negativities of youths, which led to delayed long-term youth development programs (Sechele, 2015). As a response, most authorities in southern Africa followed a preventive approach of the tertiary type albeit its inconsistencies and weaknesses. Small and

Memmo (2004) highlighted the challenges of adopting a preventive approach to youths compared to the other two contemporary approaches used in youth engagement, i.e., resilience and positive youth development. Similarly, little attention has been paid to the agency either in the development of policies or in the implementation of the same (Scchele, 2015).

Three approaches to youth unemployment challenges have been forward by Small and Memmo (2004): preventive approach, resilience approach, and positive youth development approach. Gray areas do exist between the boundaries of the three approaches, and the overall recommendation has been to strike a balance including all three approaches to develop a sustainable youth development program. Preventive approaches focus on (1) prevention of occurrence (primary type), (2) interventions in response to signs of stress (secondary type), and (3) reduction of stress by treating the problem (tertiary type). Resilience approaches focus on the critical characteristics that assist recovery from and adaptation to stressful environments. It is demonstrated under extreme stressors and manifestations of successful adaptation, despite the stressful environment. Three ways used in applying the positive youth development approaches are (1) the normal development processes, (2) promotion of programs, and (3) the organizations that feed. It is not clear how these approaches have informed the current engagement efforts and policy and resource allocation toward youth development in the SADC countries.

The use of the three youth development approaches in the engagement of youth in smart agriculture is a new approach in many countries. Engaging youth in smart agriculture reduces unemployment, secures their future interest in agriculture, and improves the image young adults have on farming.

11.6 USE OF KNOWLEDGE-INTENSIVE TECHNOLOGIES TO GENERATE YOUTH EMPLOYMENT

Knowledge and the willingness to learn are important assets for youth development. This potential builds into the overall economy at national level over time. We hypothesize that the demographic dividend from youths' focus over time is highly likely where the three development approaches are applied in youth development programs. We define demographic dividend as the accelerated economic growth that may result from a decline in a country's mortality and fertility management and the subsequent change in the age structure of the population. Rural youths can reduce the burden of risk factors by acquiring improved information skills and applying them in agricultural practices. They stand a greater chance of gaining resilience through steeling, i.e., increasing the chances of overcoming challenges; during the process of becoming future commercial farmers and agents to disseminate agricultural technical information; and when involved in e-agriculture. Youths of this era enjoy exposure to the digital world and have skills that can assist them in using information and communication technologies (ICT) to gain employment. ICT offer low-cost forms of communication to commodity markets and a range of opportunities to service the needs of the poor (Curtain, 2004). Currently it is common for e-experts to develop and use applications on both the computer and cell phones to organize and disseminate information and knowledge through short messages and videos. Knowledge is required for generating tools for good agriculture management, weather forecasting, and extension services for commodity value chains, e.g., use of drones in information generation. Such activities have received attention from universities and agricultural institutions, e.g., the Global Forum for Agricultural Innovation, the Food and Agriculture Organization of the United Nations (FAO), the International Fund for Agricultural Development (IFAD), and the Technical Center for Agricultural and Rural Cooperation (CTA) (FAO et al., 2014). The field of information management has developed fast and provides opportunities for youth employment. In short, youths are potential candidates for entrepreneurship development, leading to creation of employment in the informal sector. As a key partner in ICT, youth-led enterprises are the mobile cellular service providers. Communication using mobile devices is currently very expensive in southern Africa and network service providers can partner with youths to support lower tariffs for youth-led businesses especially agro-enterprises.

The agricultural transformation agenda has been proposed and uses the commodity value chain approach. Away from the labor-intensive traditional systems, a transformed agricultural system has emphasis on efficiency, value for money, and increased skilled labor (Ndjeunga and Bantilan, 2005). A modernized agricultural production system with mechanized operations creates confidence, advances youth skills, and could be attractive for youth. Young adults can use their skills in the operation of large agricultural machinery, thereby increasing productivity and reducing youth unemployment. In this regard, agricultural production can become an entry point in the business of youths. There is however a need to drum up support of established private large-scale commercial farmers to mentor such youths with interest in agricultural value chain for business. Similarly, local and community leaders need to support youths with long-term access to land or ownership arrangements supporting the agricultural programs for youths.

Smallholding farmers in southern Africa produce from small and fragmented farms. For those farmers in remote and poorly communicated rural communities, marketing and logistics of supplying produce and products to the market are a nightmare. The poor market linkages that exist currently define an entry point into business for youths who can provide the logistical services between the farm gate and organized markets at the district, national, or regional level. However, there will be a need to invest in accessing the national and regional commodity market databases for use in such enterprises.

11.7 MODELING YOUTH EMPLOYMENT OPPORTUNITIES IN AGRICULTURE

Current youth models are shaped by public funding usually designed to support broad government interests such as internal security, food production, and government-led infrastructure support. Although employability and skills development were often quoted as the main reason, little progress on this suggests otherwise. There has not been a deviation from the historical voluntary or mandatory national service programs, which were common in most countries in southern Africa. It is evident that more often than not the public funding mechanism has limited the commercialization drive compared to the private sector and hence the need to strengthen the public–private community partnership approach. In modeling youth engagement for business in agriculture, it is critical to consider (1) the agency's (i.e., youth) interests and capabilities (Sechele, 2015), (2) the relevant stakeholders who can support the business incubation approach, and (3) the service providers who provide the soft skills and linkages often missed by public government programs. Fig. 11.1 illustrates a model based on a crop value chain in which youths have identified three areas of interest: production, value addition and processing, and market information services using ICT.

Because of competition, it is important to emphasize here that the offer that youths will bring to the market should be of superior quality compared to others. Therefore, youth-led and youth-owned business enterprises need to be well thought out and well planned to contribute to sustained businesses. In the past a large number of youth projects did not survive the

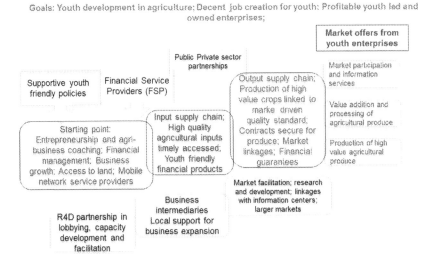

Figure 11.1 The linkages that can support the development of youth-led enterprises based on value chains that can contribute to reducing unemployment in young adults.

competition, as the services offered were not distinctively superior to the ordinary options found in the market.

Business models supported by communities and public institutions do not attract support from experienced entrepreneurs, private sector actors, and academic and financial institutions. In analyzing how the projects have been implemented in the past, it is evident that key players and opinion leaders of a particular value chain need to be brought on board sooner than later, for both viability and sustainability (Table 11.2).

The roles of some of the key players who will support agricultural youth models to improve markets and incomes are defined in the following:

1. A financial service provider (FSP) is defined here as a legal organization whose business is in financial development and management for profit. For purposes of supporting youth development programs, these organizations need to evolve from the tradition of lending institutions to design development products that include youths and other supporting partners while reducing the repayment burden on clients (youths). In other words, other guarantors need to be part of the financial instrument to reduce defaulting (DAI, 2008; The Springfield Centre, 2014).

2. An offtaker is defined here as a business leader in his/her own area who is willing to increase business by engaging small business persons who are still beginning and learning the ropes. The offtaker needs to be in good

standing financially and willing to mentor as well as act as a broker between youths in business and larger commodity markets (Balkenhol, 2014).

3. Community leadership is defined here as the heads and leaders involved in decision making for a section of the society, e.g., villages, agricultural camps/wards, provinces, and regional bodies. These leaders also shape opinion about youths and are well positioned to support and lead in the process of a positive image building process as recommended in the positive youth development approach. Access to land is one of the major limitations to the development of youth enterprises in rural communities, which is a challenge community leaders can resolve at the local level (Njenga et al., 2013).

4. Policy makers refer to people working for government agencies at different levels responsible for drafting or making representations during the process of forming new policies and revising the existing ones. This group of people shape opinions in the society and can contribute to the development of sustainable policies, especially when the private business sector players are effectively consulted during the policy formulation. In agricultural value chains, there are a range of stakeholders and associations whose specialized knowledge and experience is missed during policy formulation and wider consultation can improve the current process of developing policies related to youth development and employment. Currently, councilors in rural authorities and officials at the district and provincial levels are often left out and so are youth associations and private stakeholders in business.

5. Youth employment records and fair remuneration to avoid underemployment: Records for youth employment and development are largely difficult to find in most countries. Development of employment models for youths require collation of information at various levels. The data is very important in the analysis of progress in youth development and human capital valuation in each country. The International Labor Organization (ILO), an agency of the United Nations, has some systems in place for collecting such data, and the countries from southern Africa can learn from this (e.g., African Union, 2005).

6. Youth training facility and typologies need to be improved by using modern methods of training: training colleges and vocational institutions, including the voluntary and mandatory National Youth Service institutions, require a systematic improvement to enhance the curriculum offered and the relationship between the course content and the industrial development needs for each country in southern Africa.

Comprehensive training is fast replacing the classroom only type and traditional vocation training methods the world over and youth engagement in agricultural opportunities stand to benefit from this development. In particular the innovative involvement of private sector in youth development could increase youth employability (Fares and Puerto, 2009) and reduce the mismatch between education systems and labor market systems (YDN, 2007). Policies supporting private-owned enterprises that provide internships for youth entrepreneurs are needed to transition from supply- to demand-oriented skills development. Long-term planning of youth development program will ensure that the region benefits socioeconomically from the demographic dividend emanating as young adults become adults who are actively involved in the economy (Ashford, 2007; IMF, 2015; NYDA and VOSESA, 2011; Patel, 2009).

11.7.1 Operationalizing the Models

A forward-looking youth development agenda is imminent to reduce youth unemployment and create decent jobs for youths. Recognizing the negativities of youths who are poor, are less educated, and have AIDS (Blum, 2007), and shrinking vocational option, it is evident that youths from southern Africa need to be trained entrepreneurs. The scale required needs to directly reduce the worsening socioeconomic scenario where demand for labor outpaces supply; education disparities evident from a global job market, which requires highly skilled labor (Patel, 2009), need to be minimized. To strengthen youth entrepreneurship and job creation, the business models require that youths should invest in time to go through a complete initiation cycle. The duration will largely depend on the required expertise for a particular value chain; however, the requisite steps include

- identification of an interesting enterprise;
- increased knowledge on the input, production, output, and markets for the commodity;
- defining initial requirements for the business enterprise;
- engagement with a private sector specialist for mentoring and practical experience;
- capacity building on finances, records, and information management skills;
- linkages with FSPs, offtakers, and input markets;
- business and growth skills.

Furthermore, youths will participate in agribusinesses in a sustainable way, provided the offer that they will bring to the market will be of high quality and will be unique. We define here a market offer as any service or product, finished or intermediary that when supplied to a market fetches a value for suppliers and consumers who are willing to pay a price for it (The Springfield Centre, 2014). Market facilitation is an important determinant of success, and the Results for Development Institute (R4D) partners (Fig. 11.1) have a role to play.

Ferris et al. (2006) outlined the necessary steps required for the development of participatory agro-enterprise and a range of entry points depending on the stage at which the entrepreneurs are. For instance, groups that have no experience will focus on the organization of a functional group effort, whereas the organized groups may find entry points in the diversification and attempt testing new products in the market (Fig. 11.2). Although, in principle, markets operate on the demand and supply basis, effective marketing increases the volume of business. Marketing is defined here as a social and managerial process by which individuals and groups obtain what they need and want by creating and exchanging products and value with others (Kotler et al., 2002).

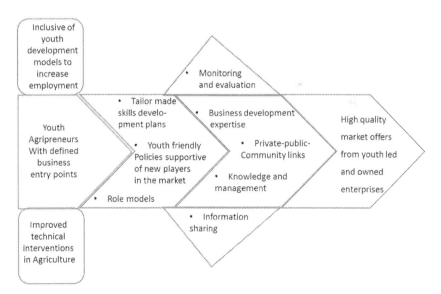

Figure 11.2 The linkages required in the development, in the establishment, and in sustaining high-quality market offers from youth-led agricultural enterprises.

11.8 CONCLUSION

A holistic program is important for youth development in southern Africa. The National Youth Service programs in their current form cannot drive employability and entrepreneurship agenda for youths in southern Africa. Youth opportunities in commodity value chains exist in production, processing, and support services. However, youth-friendly policy environment, private sector engagement at grassroots for mentoring, and improved curriculum for capacity building in transformed agriculture subsector enterprises are urgently required for youth development programs in southern Africa. Education systems that are not informed by the labor market demands will delay potential demographic dividend in many countries in southern Africa. New models for youth engagement in agriculture need to be evaluated and validated, which can reduce youth unemployment, supported by multiple stakeholders and private sector.

REFERENCES

African Development Bank, 2015. Payment for Environmental Services: A Promising Tool for Natural Resources Management in Africa. African Bank Group, Abidjan.

African Union, 2005. African Youth Charter. African Union. http://www.au.int/en/sites/default/files/treaties/7789-file-african_youth_charter.pdf.

Akoojee, S., Gewer, A., McGrath, S., 2005. Vocational Education and Training in Southern Africa A Comparative Study. Human Sciences Research Council, Cape Town.

Ashford, L.S., 2007. Africa's Youthful Population: Risk or Opportunity? USAID Population Reference Bureau, Washington, pp. 1–4. pdf.usaid.gov/pdf_docs/Pnadj952.pd.

Balkenhol, B., 2014. Microfinance, energy needs and the agricultural value chain. Value Chains in Agricultural and Green Microfinance 22.

Blum, R.W., 2007. Youth in sub-Saharan Africa. Journal of Adolescent Health 41 (3), 230–238.

Curtain, R., 2004. Creating more opportunities for young people using information and communication technology (ICT). Youth Development Journal Special Edition on Youth Employment Opportunities 48–66.

Dabalen, A., 2000. The structure of labour markets in southern Africa. In: Proceedings of the FEI/World Bank Conference on Economic Policy in Labour Surplus Economies-Strategies for Growth and Job Creation in Southern Africa May 3–5, 2000 Gaborone. Friedrich-Ebert-Stiftung/The World Bank, Gaborone.

DAI, 2008. Finance. In: Value Chain Analysis: A Synthesis Paper Micro Report No. 132. USAID, New York.

FAO, CTA, IFAD, 2014. Youth and Agriculture: Key Challenges and Concrete Solutions. Food and Agriculture Organization of the United Nations (FAO), Technical Centre for Agricultural and Rural Cooperation (CTA) and the International Fund for Agricultural Development (IFAD), Rome.

Fares, J., Puerto, S.O., 2009. Towards Compressive Training. The World Bank, Discussion paper No. 924. World Bank, Washington D.C.

Ferris, S., Kaganzi, E., Best, R., Ostertag, C., Lundy, M., Wandsschneider, T., 2006. Enabling Rural Innovation in Africa: A Market Facilitator's Guide to Participatory Agroenterprise Development. Centro Internacional de Agricultura Tropical (CIAT), Cali.

ILO, 2011. International Labour Organization. ILO, Rome. http://www.ilo.org/ilostat/faces/home/statisticaldata/conceptsdefinitions.

ILO, 2016. International Labour Organization. Labour statistics, Rome. http://www.ilo.org/ilostat.

International Monitory Fund, April 15, 2015. Sub-Saharan Africa Navigating Headwinds: Regional Economic Outlook. IMF, Washington.

Johanson, R.K., Adams, A.V., 2004. Skills Development in Sub-Saharan Africa. The World Bank, Washington, D.C.

Kanji, N., 1995. Gender, poverty and economic adjustment in Harare, Zimbabwe. Environment and Urbanization 7 (1), 37–56.

Kotler, P., Armstrong, G., Saunders, J., Wong, V., 2002. Principles of Marketing, third European ed. Prentice Hall, New Jersey.

Levinsohn, J., 2007. Two Policies to Alleviate Unemployment in South Africa. University of Michigan, Michigan, pp. 1–23.

Livingstone, I., 1985. Youth Employment and Youth Programs in Africa: Botswana. International Labour Office, Geneva.

Lough, B.J., Sherraden, M.S., 2012. Civic Service and Assert Building in Generating Livelihoods Among Youth in Africa. University of Missouri. CSD Working paper No. 12–28.

McBride, A.M., 2009. Youth Service in Comparative Perspective. Centre for Social Development Monograph No. 09-04, Washington.

McGrath, B., Brennan, M.A., Dolan, P., Barnett, R., 2009. Adolescent well-being and supporting contexts: a comparison of adolescents in Ireland and Florida. Journal of Community and Applied Social Psychology 19 (4), 299–320.

McGrath, S., 2012. Vocational education and training for development: a policy in need of a theory? International Journal of Educational Development 32 (5), 623–631.

McGrath, S., 2005. The Multiple Contexts of Vocational Education and Training in Southern Africa. Human Sciences Research Council, Cape Town.

Ndjeunga, J., Bantilan, M.C.S., 2005. Uptake of improved technologies in the semi-arid Tropics of West Africa: why is agricultural transformation lagging behind? Electronic Journal of Agricultural and Development Economics 2 (1), 85–102.

NEPAD, 2014. African Union-New Partnership for African Development. Pretoria.

Njenga, P., Mugo, F., Opiyo, R., 2013. Youth and Women Empowerment Through Agriculture. VSO Jitolee, Nairobi.

NYDA, VOSESA, 2011. How can volunteering and service promote the social and economic participation of youth in the SADC region? In: Proceedings of the Sothern African Conference on Volunteer Action for Development, October 17–19, 2011, Johannesburg.

Page, J., 2012. Youth, Jobs and Structural Change: Confronting Africa's "Employment Problem". Working Paper Series No. 155, African Development Bank, Tunis.

Patel, L., 2009. Youth development, service, and volunteering in five southern African countries. In: McBride, A.M. (Ed.), Youth Service in Comparative Perspective. Center for Social Development Monograph No. 09-04, Washington, p. 83.

SADC YearBook, 2013. Southern Africa Development Community (SADC) Statistical Year Book. http://www.sadc.int/information-services/sadc-statistics/sadc-statistics-yearbook-201/.

Sader, F. (Ed.), 2004. Youth Development Journal Special Edition on Youth Employment. Youth Development Network.

Scorgie, F., Chersich, M.F., Ntaganira, I., Gerbase, A., Lule, F., Lo, Y.R., 2012. Socio-demographic characteristics and behavioral risk factors of female sex workers in sub-saharan Africa: a systematic review. AIDS and Behavior 16 (4), 920–933.

Sechele, L., 2015. Capturing agency and voice in research: a critical review of studies of youth unemployment, self-employment and the informal sector in Botswana. Journal of Sociological Research 6 (1), 116–128.

Small, S., Memmo, M., 2004. Contemporary models of youth development and problem prevention: towards an integration of terms, concepts and models. Family Relations 53, 3–11.

The Springfield Centre, 2014. The Operational Guide for the Making Markets Work for the Poor (M4P) Approach, second ed. SDC & DFID. https://beamexchange.org/uploads/filer_public/0c/93/0c939257-39ea-4b33-b904-6eb92292b3c9/m4p_guide_management.pdf.

UNICEF, 2016. Statistical tables. In: The State of the World's Children: A Fair Chance for Every Child. UN. http://dx.doi.org/10.18356/d16f2a12-en.

Viner, R.M., Ozer, E.M., Denny, S., Marmot, M., Resnick, M., Fatusi, A., Currie, C., 2012. Adolescence and the social determinants of health. The Lancet 379 (9826), 1641–1652.

Vogel, P., 2015. Generations Jobless? Turning the Youth Unemployment Crisis into Opportunity. Palgrave Macmillan, UK.

White A.M., Gager C.T., 2007. Idle hands and empty pockets? Youth involvement in extra-curricular activities, social capital and economic status. Youth and Society 31, 75–111.

World Bank, 2015. World Bank Open Data. http://data.worldbank.org/.

World Bank, 2016. The World Bank Group. http://data.worldbank.org/indicator/SP.DYN.LE00.MA.IN/countries.

Youth Development Network, 2007. Analysis of Employment Policies in Lesotho, Malawi, Mauritius, Namibia and Tanzania as It Pertains to Youth Development. Research Report 2007.

FURTHER READING

Government of Zimbabwe, 2012. Comprehensive Agricultural Policy Framework (2012–32). Government of Zimbabwe, Harare.

International Youth Foundation, 2014. Youth Map Zambia: A Cross-Sector Analysis of Youth in Zambia. Youth Map Assessment Report. IYF, Baltimore.

Kruijssen, F., 2009. Youth Engagement in Agriculture Research. A Focus on Sub-Saharan Africa. Wageningen University and Research Center, Wageningen. 72pp.

UZ/MSU Food Security Project, 1990. In: Proceedings of the First National Consultative Workshop on Integrating Food, Nutrition and Agricultural Policy. Montclair Hotel, Juliasdale, July 15–18. Sebri Printers, Harare.

CHAPTER 12

Enabling Agricultural Transformation Through Climate Change Policy Engagement

Elias Kuntashula[1], Terence Chibwe[2], Lydia M. Chabala[1]
[1]University of Zambia (UNZA), Lusaka, Zambia; [2]International Institute of Tropical Agriculture (IITA), Southern Africa Research and Administration Hub (SARAH) Campus, Lusaka, Zambia

Contents

12.1 INTRODUCTION

Climate change effects are evident in agriculture and hence climate-smart approaches are gaining ground in southern Africa. Governments are beginning to respond by putting in place the required policies guiding investment, development, and building long-term adaptive measures in production systems. Subregional organizations and international organizations

Smart Technologies for Sustainable Smallholder Agriculture
ISBN 978-0-12-810521-4
http://dx.doi.org/10.1016/B978-0-12-810521-4.00012-8

233

including the United Nations and World Bank are among the leaders in shaping the policy discussions that are going on. Although most countries have good agricultural policies in place there are some that require attention in especially the area of institutional arrangements to capture climate change mitigation and adaptation strategies. The aim of this chapter is to discuss and evaluate the progress made in mainstreaming climate-smart approaches and strategies in agricultural policies in the Southern African Region.

Many things shape the policy discussion in the region and individual countries. The current policies and the future will depend on the projected effects of the new policy directions in assisting the citizens going forward. We define a policy as a set of overarching guidelines and principles formulated by a government to provide direction on how a particular issue such as climate change is to be managed in a country. Policy debate on critical issues such as climate threats to agriculture and natural resources should often be structured using a bottom-up approach beginning at the lowest level through the various processes until the debate in parliament. Some of the weaknesses of policies and hence limited effectiveness stem from the approach taken by government, which restricts participation by private sector and concerned citizens. Fig. 12.1 illustrates a generic framework of stakeholders that should be involved in the formulation of climate change policies. The Ministries of

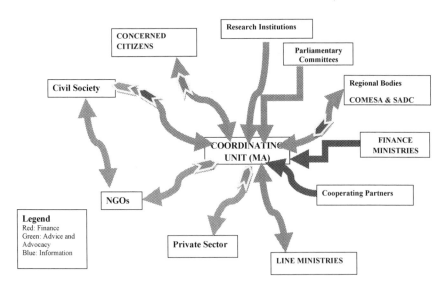

Figure 12.1 Influence network map of stakeholders in climate change formulation processes.

Agriculture and/or a unit within these ministries working in close consultation with Environmental Ministries (or Units) are responsible for developing the National Climate Change Policies and they should also be the custodian of the policies. These ministries initiate and coordinate the policy formulation processes in addition to mobilizing resources for the policy formulation and implementation. The ministries also monitor and evaluate the policy processes to take corrective steps in time. Being the lead institutions in the agricultural development and climate change policy processes, the government ministries of agriculture and environment have a lot of influence. Through their mandate, these ministries are influential in several ways. That is, they have coordination influence, decision-making influence, financial support influence, ability to attract funding information and advisory influence, formal supervisory role, and ability to exert political pressure in both agricultural and climate change policy development.

Several other key stakeholders at the national level, such as other relevant line ministries, provide input and advice in the agricultural development and climate change policy processes. For instance, the cooperating partners supplement government financing and also influence decision making on the strategic direction of the policies. They are also influential in lobbying government during the policy formulation processes. In the implementation of policies, the cooperating partners finance several programs and projects. Civil society is also influential in lobbying and advocacy to reflect the needs of vulnerable groups. Nongovernmental organizations (NGOs), which also include youth NGOs, also have influence on sensitization issues, lobbying, and advocacy. The regional bodies that include Common Market for Eastern and Southern Africa (COMESA) and South African Development Community (SADC) also influence the process by providing finances and advice to the process. They also influence decision making, something similar to what cooperating partners do. During the implementation phase, these institutions also provide technical assistance for capacity building.

An important stakeholder who often is usually forgotten in this whole setup is the concerned citizen or in the primary agricultural sense, the farmer. The farmer should be at the center of all the stages of policy formulation process. The identification of climate change problems, formulation of mitigating and adaptation policies, and their implementation should all be centered at the person "in whose reality it counts," the farmer. Other individuals and institutions influence the policy process in so many ways. For instance, research institutions undertake research and inform the policy process through their research findings. They also carry out field demonstration to test technologies, which are later used during the implementation phase.

12.2 CLIMATE CHANGE SITUATION IN THE SOUTHERN AFRICAN REGION

Climate change is one of the most widely recognized challenges to food production (Kuntashula et al., 2014). Climate change, and the effects thereof, has increasingly been receiving much attention in recent years. Eriksen et al. (2008) note that all of Africa is likely to warm during the 21st century, with the drier subtropical regions warming more than the moist tropics. They note that annual rainfall is likely to decrease throughout most of the region, with the exception of eastern Africa, where annual rainfall is projected to increase.

Southern Africa is largely an agricultural predominant region with a large proportion of the farmers coming from the smallholder sector. These mostly resource-poor farmers heavily rely on natural climatic conditions making the interannual and seasonal rainfall variability as well as changes in temperatures associated with climate change, key factors in the success of their agricultural production. Moreover, the majority of the smallholder farmers only grow the staple food crops especially cereals, hence highly exposing themselves to risks associated with changes in the climatic conditions.

Recognizing the importance that climatic hazards have on natural climatic-based agricultural production, historical knowledge of temporal changes in rainfall, temperatures, and associated extreme events is essential for successful crop production. Extreme events in the context of climate change refer to dry spell (drought) and floods among others. Wilhite and Buchanan-Smith (2005) described these extreme climate events to be among the world's costly natural disasters. With over 70% of farmers relying on rainfed agriculture in the region, any change in rainfall distribution is disastrous for crop production (Bank et al., 2007). Sivakumar (1991) reported that the success or failure of a crop depends more on the distribution of rainfall over the growing season rather than on the total rainfall in that period. Thus, many authors have attributed long dry spells within a season and shorter rain seasons as the major contributing factors to low yields (Schultz, 1974; Bank et al., 2007). Although it is not possible to influence timing and amount of rainfall (Stroosnijder, 2008), knowledge of their frequency for a variety of durations is imperative, if they are to be well managed (Nuñez et al., 2011).

The effects of climate change have never been so pronounced before in southern Africa than in the 2015–16 farming season. The region experienced an intense drought that spanned across large swathes of Zimbabwe, Malawi, Zambia, South Africa, Mozambique, Botswana, and Madagascar. The rainfall season (2015–16) has so far been the driest in more than three decades. These areas have experienced drastically delayed planting seasons, with farmers in countries such as Zambia planting the staple maize crop as late as first week of February 2016. The conditions for early crop development and pasture regrowth have been generally very poor in the entire region. The farmers that attempted planting with the first unevenly distributed rains experienced widespread crop failure. This is because the early rains were immediately followed by a prolonged dry spell that disturbed the normal plant growth. Meteorological sources such as the European Center for Medium Weather Forecasts, Climate Prediction Center, United Kingdom Meteorological Office, and International Research Institute for Climate and Society monitoring and predicting weather patterns in the region have unanimously agreed that continuation of below-average rainfall and above-average temperatures across most of the region was going to characterize the remainder of the 2015/2016 growing season. According to the WFP VAM Food Security Analysis report the rainfall forecasts for the months ahead showed a high likelihood of continued extremely difficult conditions for crop development (http://vam.wfp.org/sites/seasonal_monitor/afs/index.html). Considering that the 2014–15 season was equally poor in terms of both rainfall distribution and quantity, a scenario of extensive regional-scale crop failure and food insecurity loomed in the region. The 2014–15 agricultural season saw the regional cereal production drop by 23% due to hot and dry conditions similar to what was obtained in the 2015–16 season. The drop in cereal production is likely to be higher during the 2015/2016 season; hence increasing the region's vulnerability due to the depletion of regional cereal stocks. The number of food insecure people in the region (not including South Africa) already stood at 14 million (7.6 million in currently drought affected countries, plus 6.6 million in Democratic Republic of Congo where conflict, not drought is the driver of food insecurity, and 0.4 million in Tanzania), according to the SADC. As of early February 2016, Famine Early Warning System Network estimated that, of this total, at least 2.5 million people are in crisis (Integrated Phase Classification Phase 3) and require urgent humanitarian assistance.

Drought emergencies have been declared in most of South Africa's provinces as well as in Zimbabwe and Lesotho. Water authorities in Botswana, Swaziland, South Africa, and Namibia are limiting water usage because of low-water levels. Power outages have been occurring in Zambia and Zimbabwe as water levels at the Kariba dam have become much lower than usual.

12.3 ADOPTION OF KEY AFRICAN CLIMATE SOLUTIONS AND MAINSTREAMING OF CLIMATE CHANGE IN NATIONAL POLICIES

The adverse impacts of climate change combined with a weak adaptive capacity of most rural households in Africa necessitates the need to ascertain viable and sustainable adaptation strategies at the center of policy analysis and debate (Hisali et al., 2011).Thus, an important task facing most African countries is, therefore, to formulate policies, which facilitate the uptake of agricultural practices that increases adaptation to climate change, to ensure agricultural development and food security.

The regional and country efforts toward incorporating climate change-related policies and strategies are shown in Table 12.1. There is neither a unified policy on climate change in the southern African region nor a well spelt out climate change policy for each individual country. Some sectoral strategies can, however, be said to have taken a regional dimension. For instance, there is a climate change strategy that encompasses countries that have the membership with the regional common market grouping.

Second, countries belonging to the Southern African Customs Union have agreed to harmonize agricultural and related policies after SADC Ministers of Agriculture adopted the implementation of the Regional Agricultural Policy. Individually, not all the countries have climate change strategies, although it is recognized that those that do not have strategies are in various stages of developing them. Seven countries in the region also have National Adaptation Programmes of Action (NAPAs) as part of United Nations Framework Convention on Climate Change (UNFCCC) initiative for the least developed countries to identify priority activities that respond to their urgent and immediate needs to adapt to climate change.

A number of sectors are involved in the development of policies related to climate change depending on the governance structure in the country, for example, in Zimbabwe, Zambia, and Malawi the Ministries of Agriculture, Energy, Tourism, Mining, Natural Resources, and Environment have been involved in formulating policy guidelines in climate change issues. It should be noted that because of the cross-cutting nature of climate-smart policies there are often more than one sector or ministry

Policy engagement	Activity	Participants in the process	Framework and policies	Climate change response strategy	Output
AU-NEPAD	Key issues are enlisted by line ministries that are represented at the continental body	Member country especially lead sectoral ministries of lands, natural resources and environment, water development, ministries of agriculture and finance and other relevant sectoral ministries, International Nongovernmental Organizations, World Bank, EU	Framework is designed as an agriculture–climate change strategic tool to build capacity and address aspects of alignment, coherence, and harmonization along common and share principles and agenda on agriculture–climate change. Particularly relevant to the regional agriculture sector is CAADP, which implicitly embraces climate change issues under its strategic pillars 1 and 3. Heads of state and government Malabo declaration (2014) on accelerated agricultural growth and transformation for shared prosperity and improved livelihoods. The NEPAD international nongovernmental organization (NGO) alliance on CSA	Uses the agriculture climate change adaptation–mitigation framework to help Africa define and determine its agenda on agriculture–climate change and build informed and decisive leadership and responsibility. The two CAADP pillars strategically address some of the fundamental sources of vulnerability to climate change and vulnerability by communities in southern Africa, though their design formulation was apparently not from a climate change adaptation perspective	Enhanced capacity development in climate change issues and investment in environmental resilience programs. Africa more and better prepared to engage in global issues such as climate change that affect or are affected by agriculture

Continued

Table 12.1 Summary of climate smart agriculture technology policies at continental, regional, and country level—cont'd

Policy engagement	Activity	Participants in the process	Framework and policies	Climate change response strategy	Output
SADC	Similarly key climatic issues are enlisted by line ministries that are represented at the regional body	Member country especially lead sectoral ministries of lands, natural resources and environment, water development, ministries of agriculture and finance and other relevant sectoral ministries	Has developed a regional agricultural policy (RAP). The SADC Maputo declaration of 2003, which requires countries to commit at least 10% of their national budgets to agriculture. Although the Maseru protocol of 1996 (eventually launched in 2008), sought to establish a SADC free trade area (FTA) to enable member states to liberalize trade through removal of tariffs and nontariff barriers.	The RAP seeks to harmonize policy for agriculture and natural resources and strengthen the interventions so far guided by the SADC regional Indicative strategic development plan (RISDP) of 2003	Concerted effort in addressing climate change by regional member states
COMESA	Similarly key climatic issues are enlisted by line ministries that are represented at the regional body	Member country especially lead sectoral ministries of lands, natural resources and environment, water development, ministries of agriculture and finance, and other relevant sectoral ministries, NGOs specialized in	CAADP pillar numbers 1 and 3	All COMESA agricultural development programs are aligned to CAADP	Compliment and link with NAPs, etc.

Country (Botswana, Lesotho, Zambia, Namibia, Zimbabwe, Mozambique, Malawi, Swaziland, South Africa) specific policies	Governments formulate and put into effect and are responsible for the enforcement of the policies	Lead sectoral ministries of lands, natural resources and environment, agriculture and livestock, finance and other relevant sectoral ministries, and key stakeholder institutions	National agricultural policy (NAPs), National Agricultural Investment Plans, and nationally appropriate mitigations, NPCC in Zambia	Agricultural activities are guided by the respective national policies	Increased direct participation of in-country partners. Interaction and collaboration between government and development partners harmonized and more coherent and coordinated. Quality agriculture investment programs designed

AU-NEPAD, New Partnership for Africa's Development; *SADC*, South African Development Community; *COMESA*, Common Market for Eastern and Southern Africa. National policy informed by international and regional conventions and discourse on climate change derived from United Nations Framework Convention on Climate Change, NEPAD, SADC, and COMESA.

mandated with policy development in climate change. The challenge that may arise is the lack of capacity and/or willingness by various sectoral actors and departments to work together. This may lead to formulation of policies that may be formulated with limited awareness by other departments that may have to play a key role in the newly introduced policy. In cases where there was departmental and ministerial consultation during the policy formulation process, challenges sometimes arise when there is no further consultation in the implementation once the policy has been signed or passed by parliament. The involvement of the private sector has generally been limited. Further, a continuous consultation with the citizens or farmers who are the primary stakeholders in the whole process is also quite limited.

12.3.1 Adoption of Climate Change Policies in Individual Countries

12.3.1.1 Namibia

Agriculture in Namibia plays a critical role in the formal and informal economy supporting 70% of the population directly or indirectly through employment and income generation. It is estimated that the sector contributes over 10% to the gross domestic product (GDP). However, unlike countries such as Zambia where crops significantly contribute to the sector, it is the livestock sector in Namibia that contributes 75%–80% to the agricultural sector's GDP. The livestock industry accounts for 90% of all agricultural production in Namibia with approximately 60% of households (40% of poor households) owning cattle. The country suffers from extreme fresh water scarcity, which is compounded by increased changes in the climate. Climate change has increased daily maximum temperatures, whereas in other regions it has increased aridity due to the combined effect of variable rainfall and increased evaporation.

As noted earlier, crop production in Namibia is limited, mainly due to the arid climate and low rainfall patterns. The main rainfed crops grown include pearl millet, sorghum, and maize. The dependence on rainfed agriculture increases vulnerability of farming systems and predisposes rural households to food insecurity and poverty. In light of these challenges, the country has decided to adapt agricultural practices and increase the resilience of household livelihoods so as to sustain development and growth of the country amidst climate change.

The changes within the agricultural sector such as the number of bilateral, regional, and multilateral developments as well as current and expected impacts of climate change and agricultural land reform affecting the performance of the sector has prompted the revision of the 1995 National Agricultural Policy. This review presents a new direction in developing the sector. The reviewed policy took due cognizance of Namibia's obligations and commitment under the relevant international treaties and agreements. Although after reviewing the policy, it was taken into consideration that the African Union and its Regional Economic Communities such as SADC will follow through with a Comprehensive Africa Agriculture Development Program (CAADP). CAADP is Africa's policy framework for a growth-oriented agricultural development agenda of the African Union under the New Partnership for Africa's Development (NEPAD). This policy framework serves as a base for drafting new policies as well as positioning existing policies, laws, and regulations.

To specifically address the challenges posed by climate change to the agriculture sector and livelihoods of the Namibian people, a targeted strategic Country Climate-Smart Agriculture (CSA) program has been developed to provide a blueprint for agricultural development that enhances resilience and food security, and has mitigation cobenefits. The government developed this Country CSA Program (2015–25) through the Ministry of Environment and Tourism, Ministry of Agriculture, Water and Forestry, and Ministry of Economic Planning. The Country CSA Program (2015–25) is aligned with the AU/NEPAD CAADP, Namibia Vision 2030, National Development Plan (NDP4), Agricultural Sector Plan, and the National Climate Change Strategy and Action Plan. Inputs into the CSA program came from various stakeholders including from the relevant Namibian Ministries and Departments, agencies, civil society organizations, private sectors, researchers, and academia. Technical and financial support was provided by NEPAD Climate Change Fund, COMESA, and East African Community, and the Consultative Group for International Agricultural Research Program on Climate Change, Agriculture, and Food Security. The country CSA Program aims to build resilience of agricultural farming systems for enhanced food and nutrition security in Namibia.

Investment in climate change adaptation and mitigation programs is envisaged to come from several sources including the government, donor community, and others such as the private sector. The reliance on donors for sustainable agricultural practices that mitigate climate change may, however, affect the adoption of the practices. As recent as the second quarter of 2016, some projects promoting conservation farming in Namibia are at the verge of collapsing (Box 12.1).

Box 12.1 Conservation Agricultural in Dire Need of Funding

A desperate cry for funding went out yesterday from some 12,000 communal crop farmers and their families in the north, with the announcement that the most successful crop farming project ever in Namibia—the Namibian Conservation Agriculture Project (NCAP)—was under threat after the Cooperative League of the United States of America (CLUSA) confirmed its pilot project—worth N$2.1 million over a period of 3 years—has come to an end.

The devastating news reached crop farmers yesterday at a stage when planting of maize and mahangu is in full swing and amid extraordinarily high interest in Conservation Agriculture (CA), because of the consecutive droughts and a severe food insecurity situation, with some 600,000 Namibians in urgent need of drought food aid from the government.

Confirming the news, project coordinator of CLUSA in Namibia, Inomusa Nyati, told New Era that CLUSA's NCAP project—implemented through Creative Entrepreneurial Solutions—has been the most successful agri-project so far. NCAP 2 was established late last year in conjunction with Kongalend Financial Services, but additional funding is now of utmost importance to ensure the project continues and that trained farmers reap the benefits of the CA system.

He says NCAP 2 has the potential of reaching out to even more people, but expressed fears that securing funding will not be easy, as other organizations also have a mandate to do similar work. "What would be helpful is for NCAP to look for own funding and continue with the work we did and then invite these other projects to our field days. Results on the ground would help NCAP find new funding," he stated.

Reproduced from New Era (2016), Namibia.

12.3.1.2 Zambia

Over the past few years, Zambia has made significant progress on developing a conducive policy, strategic and institutional framework for climate change and agriculture development. The country has developed a draft National Climate Change Policy (NPCC) to provide a coordinated response to key climate change issues in the country. The vision of the policy is a prosperous climate change–resilient economy by 2030 that will have significantly increased living standards of the population and reduced its vulnerability to the impacts of climate change. Zambia also recently drafted a National Climate Change Response Strategy, which provides a good basis for a Climate Change Program. The draft strategy defines a national Goal, Vision, Objectives, and Pillars. It also updates the status of knowledge on climate change trends in Zambia and its impacts on key major sectors. It identifies clear priorities for adaptation, mitigation, and activities in cross-cutting themes and proposes a new institutional and governance structure for managing climate change issues in Zambia—the National Climate Change and Development Council. The National Agriculture Policy (NAP) 2004–15 is currently being revised to incorporate climate change-related emerging issues.

In accordance with Article 8.2(c), of the UNFCCC, Zambia has prepared its Second National Communication that identifies mitigation sectors as being the energy, agriculture, land-use change and forestry, and waste. For the energy sector—strategies identified are switching diesel fuel to biodiesel, switching from petrol to ethanol, fuel switch coal to biomass, grid extension to isolated diesel, and switching from existing isolated diesel to mini hydro. Within the agriculture sector, rural biogas, rural biomass, and conservation farming have been identified. In the land-use change and forestry, electric stoves, improved charcoal, and traditional woodstoves improved charcoal production (improved traditional, brick, metal, and charcoal retort), biomass electricity, forestation/enhancement, and sustainable agriculture are being encouraged, whereas the waste sector includes biomethanation and landfill. During the preparation of the Sixth National Development Plan, the above-mentioned projects were integrated in the plan. Furthermore, the National Climate Change Response Strategy (NCCRS) identified that the sectors associated with large greenhouse gas emissions in Zambia, including land-use (forestry and agriculture) and energy (road transport and industries (mining) primarily due to fossil fuel consumption).

With support from United Nations Development Programme (UNDP), Zambia completed a NAPA at the end of 2007. The purpose of the NAPA was to assess the impact of climate change and variability in Zambia and create a program of action for priority adaptation measures. The NAPA identified agriculture, natural resources, human health, and energy and water as priority sectors. More recently, in 2013, the country has started preparing the Nationally Appropriate Mitigation instrument that will succeed the NAPA.

To address climate change challenges, Zambian stakeholders prepared a Strategic Program for Climate Resilience (SPCR) for the Pilot Program for Climate Resilience in June 2011. SPCR is aimed at mainstreaming climate change adaptation into national plans and strategies; assists communities in highly vulnerable areas to identify and address their own climate change adaptation options as part of local development plans; incorporating climate resilience into existing community development funds to ensure their sustainability; investing in highly visible infrastructure as a way to rally public support for adaptation; building on partnerships, particularly with private sector and civil society; and strengthening the institutional foundation for a future climate change program in Zambia.

In 2010, the Ministry of Tourism, Environment, and Natural Resources developed a NCCRS, following a thorough process of stakeholder consultation. The NCCRS provides a solid basis for Zambia's Climate Change program. The strategy addresses five focal areas, namely adaptation and risk reduction, mitigation and low carbon development, cross-cutting issues, governance issues, and finance/investment framework. The strategy is still in draft form awaiting the finalization in the development of the NPCC, which is also in the final draft form. It should be noted that usually the policy would inform the strategy but the NCCRS was developed before the draft NPCC.

Mulenga (2013) showed that there had been growth in public climate finance in recent years in Zambia. The growth in public climate finance inflows averaged about 2.1% annually. Most of this growth can be attributed to improvements in government policy in relation to attracting dedicated climate finance, including the removal of a vast array of administrative barriers. This outcome reflects the increase in government's and donor's interests in climate change issues, with the understanding that the impacts of climate change can undermine development and destabilize poverty reduction strategies (Mulenga, 2013). As part of the program on mainstreaming climate change in the various government ministries, government has in the last 5 years allocated on average 7% of the national budget resources to climate change activities (see Table 12.2). The action by government to invest in climate change activities has also attracted private climate inflows.

Despite these recent improvements and growth in public climate finance, the country is still constrained in the implementation of environment, agriculture, and climate change programs. The implementation of policies has been weak primarily due to inadequate capacity (inadequate qualified and skilled technical and professional staff, financial resources and equipment).

12.3.1.3 Zimbabwe

Zimbabwe has the Comprehensive Agriculture Policy Framework 2012–32 (Government of Zimbabwe, 2012), which attempts to address the country's new challenges in the agricultural sector. The policy document does not explicitly contain climate change-related issues save for a mention under crop diversification, where breeding of drought tolerant crops is encouraged. The major policy objectives that indirectly hinge on climate change in this policy framework include ensure national and household food and

Table 12.2 Composition of the Zambian Government Allocation to Climate Change Program National Budget, 2007–12 (US$ millions)

Sector/institution	2007	2008	2009	2010	2011	2012
1. Agriculture	5%	10%	9%	10%	4%	5%
2. Environment and natural resources	23%	15%	19%	15%	20%	36%
3. Energy and water	24%	28%	10%	7%	6%	6%
4. Disaster mitigation and management unit	57%	57%	59%	41%	91%	83%
4.1 Disaster preparedness	46%	45%	52%	32%	19%	13%
4.2 Disaster response	11%	11%	7%	9%	73%	69%
Infrastructure						
A1—Ministry of Works and Supply	0%	6%	0%	1%	2%	0%
A2—Ministry of Communications and Transport	4%	4%	2%	6%	3%	1%
Weighted average	7%	10%	8%	10%	8%	7%

Allocation to mainstreamed programs per sector (US$ millions)

	2007	2008	2009	2010	2011	2012
Agriculture	10.81	15.01	18.66	18.13	7.19	17.61
Environment and natural resources	7.97	3.81	6.24	6.49	6.79	7.32
Energy and water	2.94	3.49	2.78	4.42	4.47	7.10
Disaster mitigation and management unit	0.79	0.80	0.93	0.67	10.85	11.18
Infrastructure						
A1—Ministry of Works and Supply	–	3.60	0.04	0.20	0.51	0.21
A2—Ministry of Communications and Transport	0.58	0.80	0.33	1.15	0.66	1.01
Total mainstreamed	23.09	27.52	28.98	31.06	30.74	44.42

Reproduced from Mulenga, C., 2013. Tracking of Public and Private Climate Finance Inflows to Zambia. Discussion paper prepared for the Climate Change Expert Group (CCXG) Global Forum March 19–20, 2013, Paris.

nutritional security; ensure that the existing agricultural resource base is maintained and improved; contribute to sustainable industrial development through home-grown agricultural raw materials.

However, Zimbabwe has some sectoral policies and strategies on climate change. There is a National Policy and Program for Drought Mitigation, which provides for provincial and district programs to access funding from international organizations for purposes of drought mitigation. According to Chagutah (2010) the country has also regional early warning systems and drought monitoring centers.

12.3.1.4 Malawi

In Malawi, climate change-related issues are covered in national agricultural policy and strategic plans. National Agricultural Policy (2010–16), the Food Security Policy (2006), Agriculture Sector Wide Approach (2010), National Disaster Risk Management Policy, National Land Resources Management Policy, and Strategy and National Irrigation Policy and Development Strategy of 2000 are some of the documents that include climate change adaptation related issues in Malawi (Government of Malawi, 2011, 2006). For instance, the National Agricultural Policy (NAP, 2010–16) explicitly provides strategies for climate change adaptation, such as improving vulnerability assessments to provide early warning on food security, enhancing food security, and developing community-based seed and food storage systems, improving crop and livestock production through the use of appropriate technologies, increasing resilience of food production systems to erratic rains by promoting sustainable dimba production of maize and vegetables in dambos, wetlands, and along river valleys, and developing a framework to ensure that all agriculture projects and programs undertaken in the sector have had environmental impact assessments as required by the law. In addition, the policy emphasizes strengthening the capacity of all stakeholders in issues of mainstreaming environmental management in the agricultural sector. Most of these issues are also amplified in the Agriculture Sector Wide Approach (2010) that broadly focuses on agricultural growth and poverty reduction, also specifically addresses food security and risk management and sustainable land and water management. The National Water Policy focuses on water resources management and development and recognizes the increasing incidence of droughts and floods. It calls for good catchment management to maintain/enhance ecosystems functioning and preserve biodiversity, including protection of wetlands (Mapfumo et al., 2014).

12.3.1.5 South Africa

In South Africa the Integrated Growth and Development Plan (2012) provides some guiding policies and strategies for the agricultural sector. The plan recognizes the critical challenges of climate change and clearly embraces the need for substantial public and private investments in irrigation; support of crop varieties and animal breeds that are tolerant to heat, water, and low soil fertility stresses; and imperative to build roads and marketing infrastructure to improve small farmers' access to critical inputs as well as to output markets. The South African government also has the Comprehensive Rural Development Program (CRDP), which focuses on three main pillars, namely land reform, agrarian transformation, and rural development (Government of South Africa, 2009). Vulnerabilities of the socially diverse rural communities and hence climate change adaptation processes are contextualized in the CRDP. Among the specific climate change critical developmental areas that the program tries to address include capacity building initiatives, in which rural communities are trained in technical skills, combining them with indigenous knowledge to mitigate community vulnerability to, especially climate change, soil erosion, adverse weather conditions, and natural disasters, hunger, and food insecurity; empowerment of rural communities to be self-reliant and able to take charge of their own resources; development of mitigation and adaptation strategies to reduce vulnerabilities with special reference to climate change, erosion, flooding, and other natural disasters; and increased production and sustainable use of natural resources, including related value chain development in livestock (exploring all possible species for food and economic activity) and crop farming (exploring all possible species, especially indigenous plants, for food and economic activity).

12.3.1.6 Botswana

Not until recently, has Botswana, which has been highly vulnerable to environmental and social pressures emanating from climate change for a long time, started developing a comprehensive policy response to climate change. Lately, climate change and its adverse effects on the environment, people's livelihoods, and the economy are now being taken seriously by policy makers in the country. Botswana is developing a National Climate Change Policy and Strategy and Action Plan in collaboration with the UNDP, whose objectives include (1) to develop and implement appropriate adaptation strategies and actions that will lower the vulnerability of Botswana and various sectors of the economy to the impacts of climate change; (2) to develop action and strategies for climate change mitigation; (3) to integrate climate change effectively into policies, institutional, and development frameworks in recognition of the cross-cutting nature of climate change; and (4) to ensure that Botswana is ready for the post-2015 climate regime when a new protocol applicable to all parties will be finalized.

Key policy documents such as the NDP and Vision 2016 have reflected how challenges due to climate change are to be tackled. The NDP 10 (2009–16) notes that environmental protection and climate change mitigation and adaptation need to be mainstreamed in the development processes.

12.3.1.7 Lesotho

Lesotho has had policies and strategies that had an indirect bearing on climate change issues since the 1980s. Although there was no specific policy relating to climate change, other sectoral policies and strategies addressed some of the climate issues albeit in an indirect manner. Some of the policies related to national afforestation programs, revision of agricultural policies, and formulation of environmental regulations. As climate change issues begin to take center stage across the world, the country participated and ratified many international treaties and conventions. These included the UNFCCC (1995) and Kyoto protocol (1997). Therefore, awareness campaigns on climate change issues as it relates national vulnerability were conducted in line with the ratified conventions and protocols. By 1997, the National Forestry Policy of Lesotho was adopted that directly related to climate issues with regard to forests. Forests are considered as the main resource for climate resilience because they act as sinks for carbon dioxide and other greenhouse gas emissions to the environment (Lesotho Meteorological services, 2001). Other interventions were simultaneously taking place such as the National Livestock policy, which dealt with uncontrolled grazing and environmental legislation.

Later adaptation policies and programs were documented to reduce the adverse effects of climate change. The NAPA on climate change (Government of Lesotho, 2007) prioritized adaptation options for implementation in the vulnerable areas of Lesotho. One of the key adaptation options outlined in the NAPA was capacity building and policy reform to integrate climate change in sectoral development plans (FAO, 2011). It should be noted that Lesotho like other countries in southern Africa should continuously build on the NAPA so that sectoral policies continuously align with national development and changing needs. For instance, although agriculture is recognized as one of the sectors that need to be strengthened with regard to climate change adaptation, practical actions in the local context have often been slow.

12.3.1.8 Mozambique

It is estimated that climate change in Mozambique may result in GDP decline in the range of 4%–14% (ref). Having recognized the potential adverse effects of climate change, Mozambique has had consecutive 5- or 4-year Poverty Reduction Action Plans 2006 (Wingqvist, 2011). These programs included aspects that dealt with climate change issues, although their principal focus was on enhancing agriculture and fishery productivity, employment creation, and human and social development including improving land administration and access to markets.

Following the United Nations Conference on Environment and Development summit in 1992, Mozambique gradually developed a legal framework to cope with climate change. The Ministry for Coordination of Environmental Action coordinated various working groups to determine and implement the national Clean Development mechanisms and produce the National Adaptation Program of Action (2007). Some of the adaptations actions related to low cost agricultural options took into account climate change issues so as to promote environmental sustainability.

Other international adaptation programs resulted in the mainstreaming of climate change and environmental issues in national programs. These programs included the Joint Program in Environmental Mainstreaming and Adaptation to Climate Change in Mozambique funded by Millennium Development Goals - Financing, Spain and UNDP; Mozambique Poverty and Environment Initiative funded by the government of Ireland and the Coping with Drought and Adaptation to Climate Change funded by Global Environmental Facility (Artur and Hilhorst, 2012).

It has generally been recognized that climate change and environmental issues have presented policy challenges in terms of harmonization of policies due to its multidisciplinary nature. This is because some of the climate change issues that policy seeks to address may lie outside the main coordination agencies tasked with formulation of environmental policies. Examples relate to adaptation policies in agriculture and energy that may be more explicitly outlined in their sector specific policies. Further some of the policy initiatives and programs in areas like agriculture are donor driven and sometimes bring competition for control due to resources associated with climate change programs.

12.3.1.9 Swaziland

The recognized land-use categories in Swaziland include crop agriculture, animal husbandry, forestry, extraction and collection, nature protection, settlement, and industry. It is estimated that around 10% of the total land area is suitable for agriculture production. With such a small percentage of land suitable for agriculture, the impact of climate change may have adverse effects particularly if it is associated with crop failure (Simelane and Dube, 2015), which may negatively impact household food security. With diminishing land areas suitable for cultivation, it has been recognized that policy mechanisms should be in place to ensure productivity in suitable areas (Mapfumo et al., 2015).

Although climate change issues have only received more focused and open discussion in the recent past, integration of climatic and environmental concerns into the main policies and sectors was catered for as early as the mid-1990s when the National Development Strategy and the Swaziland Environment Action Plan were formulated with the Swaziland government in 2012). As a result, climate change issues were taken into account in subsequent policies. These policies included the Comprehensive Agricultural Sector Policy (CASP), the National Food Security Policy for Swaziland (Government of Swaziland-Ministry of Agriculture and Cooperatives, 2006), National Biodiversity Conservation and Management Policy (2007), the National Biofuels Development Strategy and Action Plan (Government of Swaziland-Ministry of Natural Resources and Energy, 2008), and the draft National Energy Policy Implementation Strategy of 2009.

12.4 EFFECT OF CLIMATE CHANGE ON THE VULNERABLE GROUPS AND THEIR PREPAREDNESS (ADAPTATION AND RESILIENCE MEASURES)

There is a wide recognition that the impact of climate change is already being felt in the region, as a result of an increased incidence of disease outbreaks, flooding, and prolonged periods of drought that have resulted in rising food prices (Richards, 2008). Effects of climate change and climate variability will continue to challenge vulnerable people. Apart from mitigating the impact of climate change through reduction in greenhouse gas emissions, climate change adaptation is receiving attention as the impact of climate change is increasingly being felt. Agricultural production and access to food in the southern region are projected to be negatively impacted by changes in the timing and duration of precipitation events, daily temperatures, and levels of soil moisture. All phenomena that increase the risk of soil erosion and vegetation damage. The crop cultivar and livestock selection and breeding are dictated by climate and the choices and such as adaptation options that have to follow a cropping calendar (Below et al., 2010; Lesolle, 2012).

Thus, Smit et al. (2001) define adaptation as a process involving "an adjustment in natural or human systems in response to actual or expected climatic stimuli or their effects, which moderates harm or exploits beneficial opportunities." Adaptation may include strategies, policies, and measures that may be undertaken now or in the future to combat the impact of climate change, such as disease burden that is sensitive to climatic drivers (Ebi et al., 2005). In many instances adaptation measures are implemented reactively as a response to an actual situation that has occurred, such as observed changes in rainfall, or in other cases proactively with regard to anticipated changes. According to Below et al. (2010), historically, farmers have shown that they have been able to respond to climate variability. Coupling traditional with newly introduced adaptation practices can help farmers to cope with both current climate variability and future climate change. They note that discourse on adaptation of small-scale farmers in Africa to climate change has occurred in the absence of knowledge about existing and potential adaptation practices. This has been attributed to the fact that "current ideas about adaptation are vague, conducting focused research on potential adaptation practices and formulating appropriate advice for implementing new practices is difficult."

Farmers have thus responded to climate variability through a variety of crop management strategies, although most of these efforts may qualify more as coping than as adaptation strategies. However, the practices by the majority of farmers have gone unnoticed, undocumented, and unrefined to make them sound innovations culminating in a huge opportunity lost in many instances. Besides, these local innovations have not been challenged by agricultural intensification models that are useful in moving from subsistence to commercializing agricultural value chains. This can be attributed to the fact that farmers have long been living with climatic problems such as droughts, flooding, and within season rainfall variability, but it is only in recent years that awareness on the magnitude of the problem has been raised. This may, therefore, explain the thin practical evidence available on adaptation measures that have been pursued by farmers to date. Most of the available examples are notably related to food security, suggesting knee jerk than anticipatory or planned adaptation actions by most rural households (Mapfumo et al., 2014). For instance in Zambia farmers are practicing "Conservation Agriculture" to retain soil moisture through the use of basins and minimum disturbance to the soil as well improving soil fertility by retaining residues. Small-scale farmers also use appropriate technologies to harvest water and irrigate horticultural crops, which serve for both food security and income generation. Similarly, questions on how to take these to scale and higher intensity remain unanswered. Clearly initiatives to mitigate the effects of climate change are insufficient to guarantee sustainable food security (UN, 2012).

Programmes targeted at building long-term resilience in the smallholder farming sector are in most cases spear headed by the donor communities and their sustainability is not assured once the donors pull out. The supportive environment in terms of governments' investment in sustainable land management practices in the agricultural sector is still far beyond the rhetoric in the policy and strategic promises espoused by governments in southern Africa.

12.5 CONCLUSIONS

There are some movements by the southern African countries to undertake regional initiatives in disaster prevention, preparedness, mitigation, and response, in the wake of climate change in the region. There are, however, requirements for additional measures to reduce the impact of climate change on lives and livelihoods of affected populations. The effects of El Nino during the 2015/2016 agricultural season that have resulted in one of the driest rainfall season in over 30 years in the region calls for an accelerated pace toward enactment and implementation of climate-smart agricultural policies. It is already anticipated that the low food production will lead to higher prices of maize and other cereals, livestock, and livestock products, and in the process the region will suffer from increased hunger and malnutrition, given that approximately 70% of the region's population depend on agriculture.

There is, therefore, a need for the immediate implementation of short-, medium-, and long-term strategies in a collective and coordinated manner by the governments of the southern African region to minimize the impacts of climate change on the affected communities. Stakeholder involvement in designing instruments for implementation and continued dialog and refinement of adaptation strategies has a chance of ripping better results. In the immediate future member countries should increase budgetary allocations to disaster prevention, preparedness, mitigation, and response apart from supporting the small-scale farmers to produce the next production season. As a medium- and long-term strategy, enacting and implementation of regional climate-smart agricultural policies should be given the urgent attention. Similarly, budgetary support for adaptation and mitigation over and above the funds for agricultural intensification and development need to be part of the policy developments in southern Africa. There is a need for coordinated involvement of stakeholders, including the private sector and the farmers in all the processes of policy formulation and implementation. To achieve sustainable increases in agricultural productivity without compromising the future will take decades, therefore long-term investment funds need to be available for climate-smart agriculture programs.

REFERENCES

Artur, L., Hilhorst, D., 2012. Everyday realities of climate change adaptation in Mozambique. Global Environmental Change 22 (3), 529–536. Available from: SciVerse ScienceDirect (2012).

Bank, W., Mendelsohn, R., Hassan, R., Kurukulasuriya, P., Office, C.S., Nkonde, E., 2007. An empirical economic assessment. World 1–33.

Below, T., Artner, A., Siebert, R., Sieber, S., 2010. Micro-Level Practices to Adapt to Climate Change for African Small-Scale Farmers-A Review of Selected Literature. IFPRI discussion paper, Available from: International Food Policy Research Institute (2010).

Chagutah, T., 2010. Climate Change Vulnerability and Adaptation Preparedness in South Africa. Heinrich Böll Stiftung Southern Africa, Cape Town, South Africa.

Ebi, K.L., Smith, J.B., Burton, I. (Eds.), 2005. Integration of Public Health with Adaptation to Climate Change. Taylor and Francis, New York.

Eriksen, S., O'Brien, K., Rosentrater, L., 2008. Climate Change in Eastern and Southern Africa: Impacts, Vulnerability and Adaptation. Global Environmental Change and Human Security Report No. 2.

FAO, 2011. Strengthening Capacity for Climate Adaptation in Agriculture: Experiences and Lessons from Lesotho.

Government of Lesotho, 2007. Lesotho National Adaptation Programme of Action (NAPA) on Climate Change Under the UNFCCC. Lesotho, Maseru.

Government of Malawi, 2006. Malawi's National Adaptation Programme of Action (NAPA) Under the United Nations Framework Convention on Climate Change (UNFCCC). Department of Environmental Affairs, Ministry of Mines, Natural Resources and Environment, Lilongwe, Malawi.

Government of Malawi, 2011. Sector Policies Response to Climate Change in Malawi: A Comprehensive Gap Analysis. National Climate Change Programme, Ministry of Finance and Development Planning, Lilongwe, Malawi.

Government of South Africa, 2009. The Comprehensive Rural Development Programme Framework. Department of Rural Development and Land Reform, Pretoria, South Africa.

Government of Zimbabwe, 2012. Comprehensive Agriculture Policy Framework (2012–2032). Ministry of Agriculture, Mechanization and Irrigation Development, Harare, Zimbabwe.

GOS-MOAC, 2006. National Food Security Policy for Swaziland. Ministry of Agriculture and Cooperatives, Mbabane.

Government of Swaziland-Ministry of Natural Resources and Energy, 2008. National Biofuels Development Strategy and Action Plan (Draft). Ministry of Natural Resources and Energy, Mbabane.

Hisali, E., Birungi, P., Buyinza, F., 2011. Adaptation to climate change in Uganda: evidence from micro level data. Global Environmental Change 21 (2011), 1245–1261.

Kuntashula, E., Chabala, L.M., Mulenga, B.P., 2014. Impact of minimum tillage and crop rotation as climate change adaptation strategies on maize productivity in smallholder farming systems of Zambia. Journal of Sustainable Development 7 (4). http://dx.doi.org/10.5539/jsd.v7n4p95.

Lesotho Meteorological Services, 2001. Climate Change in Lesotho. Ministry of Natural Resources.

Lesolle, 2012. SADC Policy Paper on Climate Change: Assessing the Policy Options for SADC Member States. SADC Research and Policy Paper Series 01/2012. University of Botswana.

Mapfumo, P., Thabane, K., Mtimuni, A.M., Nkondze, M.S., Mumba, A., Sibanda, L.M., 2015. Evidence to Support Climate Change Adaptation in Lesotho, Malawi and Swaziland.

Mapfumo, P., Jalloh, A., Hachigonta, S., 2014. Review of Research and Policies for Climate Change and Adaptation in the Agriculture Sector in Southern Africa. Working paper 100. Available online: www.future-agriculture.org.

Mulenga, C., 2013. Tracking of Public and Private Climate Finance Inflows to Zambia. Discussion paper prepared for the Climate Change Expert Group (CCXG) Global Forum March 19–20, 2013, Paris.

Nuñez, J.H., Verbist, K., Wallis, J.R., Schaefer, M.G., Morales, L., Cornelis, W.M., 2011. Regional Frequency Analysis for Mapping Drought Events in North-Central Chile. Water center for arid and semiarid zones of Latin America and the Caribbean.

Richards, S.J.T., 2008. Climate Development, and Malaria. An Application of FUND. Climate Change 88 (1), 21–34.

Schultz, J., 1974. Explanatory Study to the Land Use Map of Zambia: With Special Reference to the Traditional and Semi-commercial Land Use Systems. Ministry of Rural Development, Republic of Zambia.

Simelane, Q.S.N., Dube, M.A., 2015. Swaziland Farmers Preparedness to Respond to Climate Change in Water and Land Use Practices. Available on: http://www.slideshare.net/.

Sivakumar, M.V.K., 1991. Drought Spells and Drought Frequencies in West Africa. Research Bulletin No. 13. International Crops Research Institute for the Semi-Arid Tropics (ICRISAT).

Smit, B., Pilifosova, O., Burton, I., Challenger, B., Huq, S., Klein, R., Yohe, G., 2001. Adaptation to climate change in the context of sustainable development and equity. In: McCarthy, J., et al. (Ed.), Climate Change 2001: Impacts, Adaptation, and Vulnerability. Cambridge University Press, New York.

Stroosnijder, L., 2008. Linking drought to desertification in African drylands. In: Gabriels, D., Cornelis, W., Eyletters, M., Hollebosch, P. (Eds.), Combating Desertification: Assessment, Adaptation and Mitigation strategies. Published jointly by UNESCO Chair of Eremology. Ghent University, Belgium, and Belgian Development Cooperation.

UN (Zambia), 2012. Climate Change: Barrier to Attaining Food Security. UN Policy Brief, Zambia. Available online: www.oneun.org.zm.

Wilhite, D., Buchanan-Smith, M., 2005. Drought as hazard: understanding the natural and social context. In: Wilhite, D. (Ed.), Drought and Water Crises: Science Technology and Management Issues. Taylor and Francis Group, Boca Raton.

Wingqvist, G.O., 2011. Environment and Climate Change Policy Brief – Mozambique Generic Outline.

FURTHER READING

Government of Swaziland, Swaziland Second National Communication to the United Nations Framework Convention on Climate Change (Final report).

CHAPTER 13

Integrated Assessment of Crop–Livestock Production Systems Beyond Biophysical Methods: Role of Systems Simulation Models

Patricia Masikati[1], Sabine Homann Kee-Tui[2], Katrien Descheemaeker[3], Gevious Sisito[4], Trinity Senda[4], Olivier Crespo[5], Nhamo Nhamo[6]

[1]ICRAF Zambia, Lusaka, Zambia; [2]ICRISAT, Bulawayo, Zimbabwe; [3]Wageningen University, Wageningen, The Netherlands; [4]Matopos Research Institute, Bulawayo, Zimbabwe; [5]University of Cape Town, Cape Town, South Africa; [6]International Institute of Tropical Agriculture (IITA), Southern Africa Research and Administration Hub (SARAH) Campus, Lusaka, Zambia

Contents

Smart Technologies for Sustainable Smallholder Agriculture
ISBN 978-0-12-810521-4
http://dx.doi.org/10.1016/B978-0-12-810521-4.00013-X

257

13.1 INTRODUCTION

Mixed crop–livestock production systems in sub-Saharan Africa (SSA) are changing rapidly in response to various interrelated drivers such as increased demographic pressure resulting in the growing demand for food in the form of crop and livestock products, the development of local and urban markets, and dietary preferences of urban dwellers driven by incomes (Hererro et al., 2010; Hill, 1989; Smith, 1998). These drivers in the context of climate change will continue to place increased pressure on agricultural production systems to adapt. The demand for crop and livestock products could benefit small-scale farmers if they are to take advantage of new and have access to traditional markets and are able to intensify and diversify production in a sustainable way (Jacoby, 2000; Zeller et al., 1998). This would reduce production risk, poverty, and increase resilience as farmers have diverse sources of income. Smallholder production systems have a history of low and extensive input use, as they have not developed efficient enterprises. Given that currently resources in these production systems (although limited) are being used inefficiently, as evidenced by low yields, a shift toward resilient and more sustainable systems is the key to future food security. However, mixed crop–livestock production systems are complex systems with various interacting subsystems (biophysical: crops, livestock, soil, vegetation, climate, and socioeconomic: markets, social institutions, off-farm income, and remittances local customs and policies). Climate change threatens components of these crop-livestock systems. Consequently, for research and development to have an impact on systems' efficiency, there is a need to identify the potential intervention points based on an understanding of the system's individual components and their interactions in space and time.

Simulation modeling provides a valuable framework for systems analysis of crop livestock systems. Modeling allows analysis of individual components of complex systems to understand simplistic relationships between inputs and outputs, whereas system models allow evaluation of more complex interactions and determine overall systems efficiency (de Wit, 1992).

Systems modeling has the potential to achieve many of the environmental, economic, and social objectives in which field experiments and other participatory approaches might not be able to fathom.

Such analysis has the capacity to put climate change impacts into context.

In addition, the targeting of relevant practices in mixed crop–livestock systems has always been a challenge. Hence, the importance of coupling field data and experiences with computer-based decision support tools such as simulation modeling and geographical information systems. Systems modeling tools such as the Agricultural Production Systems Simulator (APSIM) (https://www.apsim.info) can also assist in conducting ex-ante impact assessments and interactions from increased management input and increased diversity (agroecological as well as economic opportunities) along with determining efficient risk reduction strategies in the context of climate change. Modeling has been used to achieve relevant and significant interventions in farm management systems (Cabrera et al., 2008). This approach, however, has been struggling to find prominence in research for the development of relevant agricultural technologies and sustainable intensification options in smallholder farmers' decision-making processes at farm level in SSA (Carberry et al., 2003). To date, modeling has received limited attention in complex farming systems in SSA albeit the critical role they can play.

Constraints to application of this methodology are mainly lack of data (biophysical, socioeconomic), appropriate expertise, and validated modeling tools or models and low to nonparticipatory approach during model development. Increased use of these tools in Africa would significantly contribute to sharpening our understanding on impacts of different interventions and targeting of these to improve, crop–livestock productivity using much less resources as compared to field experiments (Bationo et al., 2012).

However, for a comprehensive assessment of crop–livestock systems it is imperative to develop/use modeling frameworks that can integrate and evaluate the trade-offs and synergies in the systems at different scales. The challenge is that most of these tools that are available have strength in either the crop or livestock component and would rarely include both crop and livestock or socioeconomic conditions. There is a need for an integrated framework that can be used to assess impacts of management practices at different levels of scale. Currently, there are not many tools that can do integrated assessment but the existing ones can be loosely coupled to achieve the desired outcomes. Using data from rural smallholder farming systems, the current study aims to use an integrated multimodeling approach for ex-ante impact assessment of climate change and adaptation strategies in heterogeneous communities in a particular context. First, the sensitivity of current crop and livestock production systems to climate change, and the impacts of adaptation strategies on future production systems were assessed. And then climate, crop, and livestock projections within the multidimensional trade-off analysis model (TOA-MD) were integrated to assess the economic impacts of climate change with and without adaptation strategies with the aim of assessing and producing a sustainable systems integration.

13.2 METHODOLOGY

13.2.1 Study Site

The site used for the study is Nkayi district located in Matebeland North Province in northwestern Zimbabwe. Zimbabwe is divided into five agro-ecological regions, known as natural regions, on the basis of rainfall regime, soil quality, and vegetation among other factors (Vincent and Thomas, 1961; FAO, 2006). Nkayi district is located in natural region IV, which is characterized by low annual rainfall (450–650 mm), severe dry spells during the rainy season, and frequent seasonal droughts (FAO, 2006). The area is also characterized by semiextensive mixed crop and livestock farming systems. Predominant soils in the area are Kalahari sands, which are low in plant nutrients especially N, P, and S and cation exchange capacity owing to low clay and organic matter contents (Grant, 1967; Nyamapfene, 1991).

Crop and livestock enterprises are complementary and at the same time competitive. Livestock are a source of draft power, cattle manure used as organic fertilizer, milk, and cash income. On the other hand, crop residues are fed to livestock. Due to increasing demographic pressure and demand for food, farmers are forced to extend cropping activities to marginal lands, rangelands, and forest areas resulting in livestock marginalization, reduced fallow periods, and ecological degradation (Abegaz, 2005; Muhr, 1998; Powell et al., 2004). The district has higher livestock numbers and poverty levels are also higher as compared to other districts in the same natural region (Homann et al., 2007; ZimVAC, 2013). There is great potential to increase both crop and livestock production, household income, and poverty reduction. Current maize yields are uneconomically low ranging between 300 and 500 kg/ha, milk production is also low, i.e., 1–1.5 L/day/cow and livestock mortality rates greater than 17% (Homann et al., 2007; Masikati, 2011). This is against the background of attainable maize yields of greater than 3 t/ha. Low livestock and crop production under smallholder farms are mainly caused by suboptimal performance related to access to technology, markets, and management aspects rather than to low physical potential (Cai and Rosegrant, 2003; Rockström et al., 2003).

There are a number of management practices that are available but the challenge is targeting relevant interventions to different farm typologies and assess trade-offs and benefits in terms of crop and livestock production and also impacts on socioeconomic conditions at the household level.

13.2.2 Integrated Assessment

The biophysical and economic models operated externally and exclusively to generate data from a range of long-term scenarios and their costs and benefits. The integrated system analysis was used to bring together data obtained from the biophysical and economic models and assist to determine profitable management practices that can increase on-farm production and resilience across existing farming typologies. This assisted in identifying management practices that can address current farming systems constraints driven by climate change and soil degradation and facilitate development along sustainable pathways. Management practices will also be evaluated for their contribution toward food, soil fertility, feed, household income, and poverty reduction. The approach took into account the heterogeneity of entire farm populations and also integrated the impacts of projected economic development. Here, three farm typologies mainly defined by livestock ownership were considered. Table 13.1 shows other main characteristic of these farms types.

13.2.3 Crop Model Description and Parameterization

Crop models such as the APSIM have been developed to simulate biophysical processes in farming systems in relation to the economic and ecological outcomes of management practices in current or future farming systems (McCown et al., 1996; Jones and Thornton, 2003; Steduto et al., 2009). APSIM is a modeling tool that is used worldwide for developing interventions targeted at improving farming systems under a wide range of management systems and conditions (Whitbread et al., 2010). The model has been used extensively in Africa, for example, in Zimbabwe to assess impacts of maize–mucuna rotations on maize production and soil water and nutrient dynamics (Masikati et al., 2014), and impact of climate change in maize production systems, Zimbabwe (Rurinda et al., 2015). In the Sahel Akponikpe et al. (2010) investigated millet response to N with a view to establish recommendations for N application better adapted to smallholder farmers. Delve et al. (2009) evaluated P response in annual crops in eastern and western Kenya. Ncube et al. (2008) assessed the impact of grain legumes on cereal

Table 13.1 Based on system characteristics of 160 mixed farms used for the analysis, by farm type, in Nkayi district

Variables	Units	0 Cattle Mean	1–8 Cattle Mean	>8 Cattle Mean	Total Mean	Total Std. Dev.
Proportion in community	%	42.5	38.1	19.4		2.5
Household members	People	5.9	6.9	7.4	6.6	
Proportion of female headed households	%	27.9	31.1	22.6	28.1	
Net returns maize	US$/farm	60	162	63	100	121
Net returns other crops	US$/farm	31	62	35	44	53
Net returns cattle	US$/farm	0	472	1347	443	586
Net returns other livestock	US$/farm	9	19	15	14	29
Off-farm income	US$/farm	220	300	294	265	217
Farms with maize	%	98.5	100.0	100.0	100.0	0.1
Maize area	Ha	1.1	1.4	1.8	1.3	0.8
Maize grain yield	kg/ha	497	826	675	657	531
Farms with small grains	%	23.5	32.8	41.9	30.6	46.2
Small grain area	Ha	0.7	0.7	1.0	0.8	0.8
Small grain yield	kg/ha	393	726	327	512	622
Farms with legumes	%	33.8	49.2	48.4	42.5	49.6
Legume area	ha	0.4	0.4	0.5	0.4	0.3
Legume yields	kg/ha	452	722	388	557	541
Cattle[a]	TLU	0	5.4	13.9	4.7	4.7
Other livestock[a]	TLU	0.3	0.5	1.6	0.6	0.9

[a]Herd size: Cattle = 1.14 Tropical Livestock Unit (TLU), donkeys = 0.5 TLU, goats, and sheep = 0.11 TLU.

crops grown in rotation in nutrient-deficient systems in Zimbabwe. Shamudzarira (2003) evaluated the potential of mucuna green manure technologies to improve soil fertility and crop production in southern Africa, whereas Robertson et al. (2005) evaluated the response of maize to previous mucuna and N application in Malawi. For the study site, the model has been calibrated (Masikati et al., 2014) and can be used with confidence in conducting ex-ante analysis of alternative management strategies aimed at improving systems productivity.

13.2.4 Livestock Model Description and Parameterization

The Livestock Simulator (LIVSIM) (Rufino, 2008) simulates pasture and animal production using a series of mathematical equations based on various biophysical inputs that control or influence production. The model simulates livestock production using a monthly time step, based on specific genetic potential and feed intake, this is done by following the concepts of Konandreas and Anderson (1982) and taking into consideration specific rules for herd management. Protein and energy requirements are calculated on derivations by the Agricultural and Food Research Council (AFRC) (1993), whereas actual feed intake is simulated according to Conrad (1966). The model can be used to measure the way in which the numbers and classes of livestock on a property drive the total demand for pasture, the match (or mismatch) between the supply of and demand for pasture, and overall animal production of beneficial products and services. The LIVSIM model has been evaluated for livestock production in natural region II and III in Zimbabwe by Rufino (2008) and Rufino et al. (2011) and also evaluated satisfactorily for the study area. We aim to use the model to evaluate different combinations of management practices on production levels under different feeding regimes.

13.2.5 Economic Model

For the analysis of the current and alternative mixed systems we used the Tradeoff Analysis model for TOA-MD as "regional" economic model (Antle, 2011). The TOA-MD model is a parsimonious, generic model for analysis of technology adoption and impact assessment (Claessens et al., 2009), ecosystem services analysis (Antle and Valdivia, 2011), and climate change and adaptation of impact assessment (Claessens et al., 2012). The TOA-MD model simulates technology adoption and impact in a population of heterogeneous farms. Farms are assumed to be economically rational and to choose between systems based on expected economic returns. The simulation model uses data on the spatial variability in economic returns to represent heterogeneity in the farm population.

13.2.6 Stakeholder Engagement

A participatory approach with stakeholders at different levels was used to develop representative agricultural pathways (RAPs). RAPs are a set of plausible future scenarios that include both biophysical and economic changes that are envisaged in the future for a particular location. They provide parameters for long-term projections of economic development and also serve as a source of reference for predictions that affect small-holder farmers taking into account both biophysical and socioeconomic constraints. Involving stakeholders in the development of RAPs is important as these will reflect reality, acceptable, and high possibility of adoption. As clearly said by Dorward et al. (2003) low adoption of technologies can be attributed to lack of stakeholder participation in the development of technologies and lack of considering market accessibility and incentives.

13.2.7 Climate Data and Crop–Livestock management

Simulations were run for 30 years from 1980 to 2010 using daily climate data (rainfall, minimum and maximum temperatures, and solar radiation) obtained from the Meteorology Department (Nkayi district). Sandy soils predominant in Nkayi district were used for simulation, soil physical, and chemical characteristics are described in Table 13.2. A short duration Maize variety SC401, sorghum variety *Macia*, mucuna, and groundnut variety *Chico* was planted at

Table 13.2 Soil initial conditions used for calibration of Agricultural Production Systems Simulator (APSIM) and Decision Support System for Agrotechnology Transfer (DSSAT) crop models. Soil samples were collected from experimental sites in December 2008 from the Nkayi district

Parameter	Soil layer (cm)					
	0–15	15–30	30–45	45–60	60–75	75–100
Organic carbon (%)	0.52	0.43	0.35	0.30	0.21	0.21
[a]NO_3-N (ppm)	3.08	2.16	2.30	2.21	2.55	1.07
Airdry (mm/mm)	0.03	0.07	0.09	0.09	0.09	0.09
[a]LL 15 (mm/mm)	0.06	0.10	0.13	0.13	0.18	0.22
[a]DUL (mm/mm)	0.16	0.18	0.19	0.20	0.22	0.24
[a]SAT (mm/mm)	0.41	0.41	0.41	0.37	0.36	0.34
Bulk density (g cm^{-3})	1.43	1.42	1.42	1.55	1.55	1.61

[a]NO_3-N = Nitrate-nitrogen, LL 15 = Crop Lower Limit, DUL = Drained Upper Limit, SAT = Saturation.
Masikati, 2011.

varying planting densities. Primary survey data from existing projects (ICRISAT CPWF, SLP studies), together with secondary data, were used to characterize the base systems. Crop and livestock models, together with climate data, were used to simulate impacts of different management practices on systems' productivity and economic returns to the farm populations. Through participatory approaches such as innovation platforms and participatory modeling exercises, realistic alternative technologies (Table 13.3) obtained for the crop–livestock systems were assessed for their potential to improve the systems in general and as adaptation strategies to climate change.

Livestock largely depended on natural pasture for feed with adjuncts such as crop residues and locally available tree pods during the dry season. During the rainy season, animals have enough feed from natural pasture but as the season progresses, feed quantity and quality reduce substantially (Fig. 13.1). Farmers in Nkayi indicated that feed shortages occur mainly during the dry season starting from August to November. Peak months for feed shortages are September and October (Masikati, 2011). In this study, crop residues from different crop management systems will be fed to livestock in particular cattle during the dry season. The LIVSIM model will be used to assess impacts of different feeding regimes on livestock milk production, offtake, and mortality.

13.3 RESULTS

13.3.1 Sensitivity of Current Crop–Livestock Production Systems to Climate Change and Impacts of Adaptation

Results show that impacts of climate change will be varied across farm categories and crops. Average impacts on maize grain will be 9%, 3%, and −2%, whereas that for stover will be −5%, −6%, and −4% for farmers with no cattle, small herds, and large herds, respectively. Both sorghum and groundnuts grain and stover will not be negatively affected, average future yields show increases >12%, mucuna biomass also on average will not be negatively affected. Although crop production will not be substantially impacted by climate change, low input systems' yields will be very low and farmers will continue to have annual insufficient maize grain for consumption (Table 13.1). Although farmers with cattle are currently food secure, production efficiency is very low as average maize production is way below potential of the area. Improving management will substantially improve crop production; however, without adaptation strategies production will be negatively affected. Adapting to climate change will substantially reduce impacts of climate change on maize except stover (−5%) under the no-cattle farmer category. Generally maize grain with adaptation will be

Table 13.3 Current and alternative technologies evaluated for crop livestock systems improvement across three household types

Farm type	Crop current system	Crop future system	Area sizes current system	Area sizes future system	Management current system	Management future system
0 cattle	Maize	Maize			Short duration hybrids plus retained OPV seeds, no fertilizer, no manure application, 2.4 plants/m²	Medium duration hybrids, 60% maize under rotation (30% legume residue retention), 20 kg N/ha three plants/m²
	Groundnuts	Groundnuts			Retained seeds, No fertilizer, No manure, 4 plants/m²	Improved variety, 20 kg P/ha, 6 plants/m²
1–8 cattle	Maize	Maize			Short duration hybrids plus retained OPV seeds, 6 kg N/ha, 135 kg/ha manure application, 2.4 plants/m²	Medium duration hybrids, maize under rotation (30% legume residue retention), 30 kg N/ha, 1100 kg/ha manure application, 3 plants/m²
	Sorghum	Sorghum			Retained OPV seeds, No fertilizer, No manure, 4 plants/m²	High yielding dual purpose medium duration, 100% sorghum under rotation (30% legume residue retention), 20 kg N/ha, 8 plants/m²
	Groundnuts	Groundnuts			Retained seeds, No fertilizer, No manure, 4 plants/m²	Improved variety, 20 kg P/ha, 6 plants/m²
		Mucuma (forage legume)				Improved variety 8 plants/m² (getting residual P from manure)
>8 cattle	Maize	Maize			Short duration hybrids, 10 kg N/ha, 560 kg/ha manure application, 2.5 plants/m²	Medium duration hybrids, maize under rotation (30% legume residue retention), 30 kg N/ha, 3215 kg/ha manure application, 3 plants/m²
	Sorghum	Sorghum			Retained OPV seeds, no fertilizer, no manure, 4 plants/m²	High yielding dual purpose medium duration, 100% sorghum under rotation (30% legume residue retention), 20 kg N/ha, 8 plants/m²
	Groundnuts	Groundnuts			Retained seeds, No fertilizer, No manure, 4 plants/m²	Improved variety, 20 kg P/ha, 6 plants/m²
		Mucuna				Improved variety 8 plants/m²

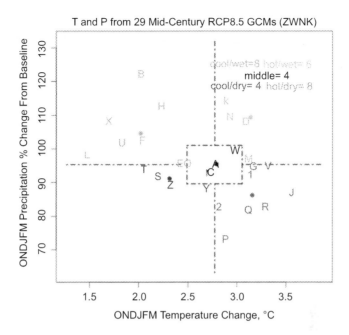

Figure 13.1 Mid-century temperature and precipitation changes in Nkayi, Zimbabwe, from 29 general circulation models (GCMs) under representative concentration pathway (RCP8.5). GCMs are represented by different letters with different dots showing average changes in precipitation and temperature as predicted by GCMs showing projected cool–wet/cool–dry/hot–wet/hot–dry, and middle conditions.

increased by 23%, 22%, and 47% for no–cattle, small herds, and large herds, respectively, under adaptation 2 (maize in rotation with legumes). With adaptation both groundnuts and sorghum show positive yield increases and lower variation for groundnuts.

13.3.2 Sensitivity of Current Livestock Production Systems to Climate Change and Impacts of Adaptation

Range and productivity are expected to decline in the future climate (Descheemaeker et al., 2016), which will in turn affect livestock productivity. Offtake is low for both small–herd and large–herd households in the current climate (Fig. 13.2). It will even become lower in the future climate but when an adaptation package is used (low rate fertilizer application on medium duration maize, maize–mucuna rotation, and the decline in offtake is minimized). Reduced grass and crop residue availability lower feed intake, resulting in up to 40% milk reduction, and 50% less offtake. The adaptation package improved feed supply, which nearly offset climate change impacts (Fig. 13.3).

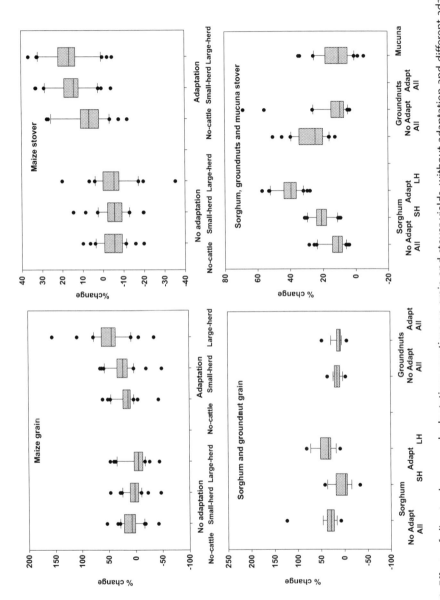

Figure 13.2 Effects of climate change and adaptations practices on grain and stover yields, without adaptation and different adaptation options across farm categories.

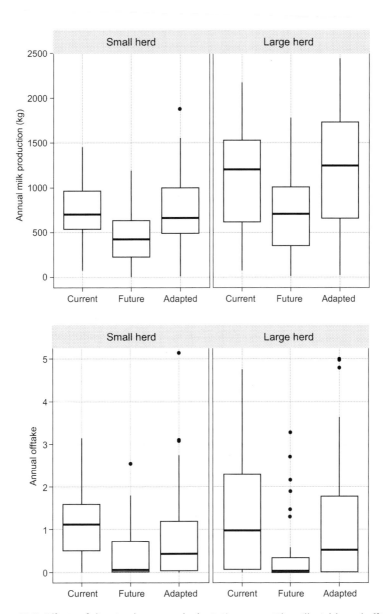

Figure 13.3 Effects of climate change and adaptations on cattle milk yields and offtake, current, future without and with adaptation, for farmers with different herd sizes.

13.3.3 Vulnerability and Benefits From Climate Change Adaptation

Even if impacts of climate change would be moderate on crops and livestock, more than half the farming population would be negatively affected and hence exposed to greater vulnerability, 45% of those without cattle, 61% and 71% of those with small and large herds, respectively, affected by feed availability. Net impacts are small for farms without livestock (3%), and much larger for farms with large herds (23%). Even though those with cattle face higher absolute losses, those without cattle are much poorer, and exposed to greater vulnerability than before climate change. The tested adaptations benefited most farms and enhanced their resilience. Farms without cattle would gain more in relation to their base income, but not in absolute terms, because these interventions would improve crop productivity more than livestock productivity Table 13.4.

13.4 DISCUSSION

13.4.1 Temperature and Rainfall Variability

In Nkayi district, changes in historical climate show increasing temperature trends and future climate temperatures are projected to increase by approximately 1–2°C in the near future, 2–3°C in the mid-century, and 2–5°C by the end of century. Projections with medium confidence show that rainfall change direction and amplitude are uncertain, yet averages would remain within or close to baseline variability, seasonality seems to remain unchanged with possible monthly rainfall reduction at the beginning of the rainy season. These projected changes will have different impacts on plant production and crop yield as production may increase or reduce depending on plant response to interactions between the different aspects of climate (carbon dioxide, temperature and precipitation).

Combinations of drought (as a result of rainfall variability) and high temperatures have adverse effect on the growth and development of both crops and the veld. Drastic yield reduction (in most cases) or total crop failure often occurs especially when drought conditions coincides with reproductive stages of the crop phenology (Schmidhuber and Tubiello, 2007). In cereals for instance, lack of thermo-tolerance and failure to maintain water relation during both pick water shortage and high temperature periods have been reported to disrupt physiological growth process leading to crop loss (Machado and Paulsen, 2001; Reddy et al., 2004).

Table 13.4 The relationship between climate impacts on cropping systems in Nkayi and related adoption of adaptations to climate extremes

Stratum	Adaptation	Climate impacts on net returns					Adoption of adaptations	
		Vulnerability	Gains	Losses	Net impact	Resilience	Adoption rate	Adopter gains
No cattle	None	45	28	−25	3	91	n.a.	n.a.
No cattle	Adapted	1	139	19	119	100	96	136
Small herd	None	61	32	−41	−9	79	n.a.	n.a.
Small herd	Adapted	25	51	12	24	100	98	51
Large herd	None	71	34	−57	−23	79	n.a.	n.a.
Large herd	Adapted	42	48	6	8	100	80	87

Selection of suitable crop cultivars with drought tolerance and the capacity for drought avoidance can also reduce the impact of extreme weather events on crops. Yue et al. (2006), Touchette et al. (2007), and Hurd (1974) reported on both tolerance and avoidance as it relates to maintenance photosynthetic efficiency and resistance to loss of moisture from the plant. Drought tolerance is, therefore, a cultivar specific trait, which breeders need to work on as more work on adaptation to climate change develops. At the point when plants are about to lose photosynthetic capacity only irrigation can change the outcome and often this is out of reach for smallholder farmers. However, tolerance also implies the capacity of the crop to keep photosynthesis rates at very reduced levels during the stress period. There is more work required on both drought and decication tolerance on commonly grown crops in southern Africa.

Increases in carbon dioxide can increase net photosynthesis rate and at the same time reduces the stomatal conductance leading to reduced respiration (Asseng et al., 2015). Responses would vary by crop species, for example, C4 and C3 plants when CO_2 is increased to 500–550 ppm grain yield can be increased by 10%–20% and by <13% for C3 and C4, respectively (Asseng et al., 2015). However, impacts of elevated CO_2 on plant production depend on other factors such as nutrient and water availability. Stomatal closure in response to high CO_2 concentrations and drought have been discussed widely as a mechanism by which plants tolerate or avoid stress (Flexas and Medrano, 2002; Medrano et al., 2002). Responses that have been reported so far at global level might not reflect similar impacts at farm level due to limiting factors across heterogeneous farms. Impacts will be varied across smallholder farming systems livelihoods; hence imperative to understand the impacts for planning and development of adaptation strategies.

13.4.2 Herd Sizes and Soil Fertility Management

The current study shows that both crop and livestock production will be negatively affected by climate change with no adaptation. However, impacts vary across farm categories and this is mainly attributed to differences in management. Farmers with larger herds (better-off) tend to better manage their farms; hence improved fertility. In this study, soils under the better-off farmers had higher initial N and organic carbon, this improves yields; hence maize grain sufficiency is higher (positive) for those with cattle than those who do not have (−13% and −19% current and future, respectively). Impact of soil characteristics is clearly shown, soils from farmers with larger herds have better capacity to buffer impacts of climate with adaptation than the other soils. Folberth et al. (2016) show that without fertilizer soil type related yield variability generally outweighs the simulated interannual variability in yield due to weather. It is, therefore, imperative to manage soils so that they can buffer and not amplify impacts.

13.4.3 Livestock Feed

Livestock play an important role for different products and services and for diversification of livelihoods in Nkayi. Livestock simulations show that animals will be affected due to possible reduction in biomass production. In this study we only assessed impacts of climate change on livestock in-terms of feed availability; hence results show negative impacts without adaptation. However, with adaptation there is potential to offset impacts of climate change on livestock production. Reductions of the impacts come about by improving feed production on-farm. Generally rangelands are degraded and livestock continue to be marginalized as more land is converted to crop production. Improving on-farm livestock feed has both environmental and economic befits; hence sustainable pathways for livestock production in future systems.

Farming systems in Nkayi are highly heterogeneous agro-ecologically and economically; in the current study we simulated impacts of climate change on three farm types. Although all will be impacted, those with no cattle will be more negatively impacted than those with livestock and are growing diverse crops. In such heterogeneous systems, it is imperative to use approaches that can assess impacts in context, as this enables researchers to explore the nature and implications of variations across farm type (Coe et al., 2016; Vanlauwe et al., 2016). Farm heterogeneity has been a major socioeconomic indicators for potential adaptation and progress beyond biophysical limitations on smallholder farms. Work of Chikowo et al. (2014) has shown that the potential for the least resource-endowed farmers typologies to intensify agricultural production is very limited and often remittance (off-farm income) are important sources for livelihoods in these communities. With climate change the farms with low-to-no resources including cattle become more exposed to threat from impact of climate variability.

13.4.4 Systems Approach

The integrated assessment approach is an important tool that can be used to assess both biophysical and socioeconomic impacts of climate change in heterogeneous farming systems. The approach also includes participatory development of representative agricultural pathways with stakeholder, making adaptations packages tailored to different farm types. Stakeholders define the technologies that will be evaluated, as well as the changes needed in the institutional and policy environment to facilitate large-scale uptake of these technologies. This can inform the design of more transformative solutions, including policies, institutional arrangements, and social organization, toward solving the complex issues in farming under changing climate. In rainfed systems, agricultural technologies always bring about varying results and uptake has always been a challenge due to inconclusive trade-offs analyses across environments and socioeconomic conditions. Use of participatory approaches coupled with computer-based integrated assessments can achieve many of the environmental, economic, and social objectives, where field experiments and other participatory approaches might not be able to fathom.

13.5 CONCLUSION

Due to complexity and heterogeneity in crop–livestock systems, agricultural systems models are often used to understand the impact of climate change and to assist in development of adaptation strategies. There are projected impacts of climate change on agricultural production; however, there is paucity information about the extent to which climate change impacts will translate into reductions in human welfare. The multimodeling regional integrated assessment approach that we used showed both impacts on crops–livestock systems and also on socioeconomic impacts. Adoption of most adaptation strategies is usually hampered by aspects such as access to markets, to credit facilities, supportive policies, and others, the participatory approach that we use in developing representative agricultural pathways facilitates potential adoption and development of information that can be used for informed decision making or development of country-specific adaptation strategies.

Use of improved cultivar, increased use, and management of soil fertility inputs, better livestock feed systems and integration of crop–livestock and people-oriented approaches stand a high chance of reducing impacts of climate variability and change. There is need for more work on refining modeling approaches, which combine biophysical and socioeconomic analysis in southern Africa in light of climate change.

REFERENCES

Abegaz, A., 2005. Farm management in mixed crop-livestock systems in the Northern Highlands of Ethiopia. Tropical Resource Management Papers, No 70, ISSN 0926–9495.

Agricultural and Food Research Council (AFRC), 1993. Energy and Protein Requirements of Ruminants. An Advisory Manual Prepared by the AFRC Technical Committee on Response to Nutrients. CAB International, Wallingford, UK.

Akponikpe, I.P.B., Gerard, B., Michels, K., Beilders, C., 2010. Use of the APSIM model in longterm simulation to support decision making regarding nitrogen management for pearl millet in the Sahel. European Journal of Agronomy 32, 144–154.

Antle, J., 2011. Parsimonious multi-dimensional impact assessment. American Journal Agricultural Economics 93 (5), 1292–1311.

Antle, J.M., Valdivia, R.O., 2011. TOA-MD 5.0: Trade-off Analysis Model for Multi-Dimensional Impact Assessment. Available at: http://trade-offs.oregonstate.edu. Accessed on 21 March 2014.

Asseng, S., Zhu, Y., Wang, E., Zhang, W., 2015. Crop modeling for climate change impact and adaptation. In: Crop Physiology. Academic Press, San Diego, pp. 505–546.

Bationo, A., Tabo, R., Kihara, J., Hoogenboom, G., Traore, P.C., Boote, K.J., Jones, J.W., 2012. Building capacity for modeling in Africa. In: Kihara, J., Fatondji, D., Jones, J.W., Hoogenboom, G., Tabo, R., Bationo, A. (Eds.), Improving Soil Fertility Recommendations in Africa Using the Decision Support System for Agrotechnology Transfer (DSSAT). Springer, Netherlands, pp. 1–7.

Cabrera, V.E., Breuer, N.E., Hildebrand, P.E., 2008. Participatory modelling in dary farm systems: a method for building consensual environmental sustainability using seasonalclimate forecasts. Climate Change. 89, 395–409. http://dx.doi.org/10.1007/s10584-007-9371-z.

Cai, X., Rosegrant, M.W., 2003. World water productivity: Current situation and future options. In: Kijne, J.W., Barker, R., Molden, D. (Eds.), Water Productivity in Agriculture: Limits and Opportunities for Improvements. CAB International, pp. 163–178.

Carberry, P., Gladwin, C., Twomlow, S., 2003. Linking simulation modeling to participatory research in smallholder farming systems. In: Delve, R.J., Probert, M.E. (Eds.), Modelling Nutrient Management in Tropical Cropping Systems. ACIAR Proceedings No. 114, pp. 32–46.

Claessens, L., Stoorvogel, J.J., Antle, J.M., 2009. Ex ante assessment of dual-purpose sweet potato in the crop-livestock system of western Kenya: a minimum-data approach. Agricultural Systems 99 (1), 13–22.

Claessens, L., Antle, J., Stoorvogel, J., Valdivia, R., Thornton, P., Herrero, M., 2012. A method for evaluating climate change adaptation strategies for small-scale farmers using survey, experimental and modeled data. Agricultural Systems 111, 85–95.

Chikowo, R., Zingore, S., Snapp, S., Johnston, A., 2014. Farm typologies, soil fertility variability and nutrient management in smallholder farming in Sub-Saharan Africa. Nutrient Cycling in Agroecosystems 100 (1), 1–18.

Coe, R., Njoloma, J., Sinclair, F., 2016. Loading the dice in favour of the farmer: reducing the risk of adopting agronomic innovations. Experimental Agriculture 1–17.

Conrad, H.R., 1966. Symposium on factors influencing voluntary intake of herbage by ruminants – physiological and physical factors limiting feed intake. Journal of Animal Science 25, 227–243.

de Wit, C.D., 1992. Resource use efficiency in agriculture. Agricultural Systems 40 (1), 125–151.

Descheemaeker, K., Oosting, S.J., Tui, S.H.K., Masikati, P., Falconnier, G.N., Giller, K.E., 2016. Climate change adaptation and mitigation in smallholder crop-livestock systems in sub-Saharan Africa: a call for integrated impact assessments. Regional Environmental Change 1–13.

Delve, R.J., Probert, M.E., Cobo, J.G., Ricaurte, J., Rivera, M., Barrios, E., Rao, I.M., 2009. Simulating phosphorus response in annual crops using APSIM: model evaluation on contrasting soil types. Nutrient Cycling in Agroecosystems 84 (3), 293–306.

Dorward, A., Poole, N., Morrison, J., Kydd, J., Urey, I., 2003. Markets, institutions and technology: missing links in livelihood analysis. Development and Policy Review 21 (3), 319–332.

FAO, 2006. Fertilizer use by crop in Zimbabwe. Land and plant nutrition management service. Land and water development division. First version, published by FAO, Rome Famine Early Warning Systems Network (FEWS NET) (June 2009) Zimbabwe food security. Update http://www.fews.net/docs/Publications/Zimbabwe_FSU_June2009.pdf. 16/08/2010.

Flexas, J., Medrano, H., 2002. Drought-inhibition of photosynthesis in C3 plants: stomatal and non-stomatal limitations revisited. Annals of Botany 89 (2), 183–189.

Folberth, C., Skalský, R., Moltchanova, E., Balkovič, J., Azevedo, L.B., Obersteiner, M., Van Der Velde, M., 2016. Uncertainty in soil data can outweigh climate impact signals in global crop yield simulations. Nature communications 7.

Grant, P.M., 1967. The fertility of sandveld soil under continuous cultivation. Part II: the effect of manure and nitrogen on the base status of the soil, Rhodesia, Zambia Malawi. Journal of Agricultural Research 5, 117–128.

Herrero, M., Thornton, P.K., Notenbaert, A.M., Wood, S., Msangi, S., Freeman, H.A., Bossio, D., Dixon, J., Peters, M., Steeg, J., Lynam, J., 2010. Smart investments in sustainable food production: revisiting mixed crop-livestock systems. Science 237, 822.

Hill, A.G., 1989. Demographic responses to food shortages in the Sahel. Population and Development Review 15, 168–192.

Homann, S., van Rooyen, A., Moyo, T., Nengomasha, Z., 2007. Goat production and marketing: Baseline information for semi-arid Zimbabwe. International Crops Research Institute for the Semi-Arid Tropics, pp. 84.

Hurd, E.A., 1974. Phenotype and drought tolerance in wheat. Agricultural Meteorology 14 (1–2), 39–55.

Jacoby, H.G., 2000. Access to markets and the benefits of rural roads. The Economic Journal 110 (465), 713–737.

Jones, P.G., Thornton, P.K., 2003. The potential impacts of climate change in tropical agriculture: the case of maize in Africa and Latin America in 2055. Global Environmental Change 13, 51–59.

Konandreas, P.A., Anderson, F.M., 1982. Cattle Herd Dynamics: An Integer and Stochastic Model for Evaluating Production Alternatives. ILCA Research Report 2, ILCA Publications, Addis Ababa.

Machado, S., Paulsen, G.M., 2001. Combined effects of drought and high temperature on water relations of wheat and sorghum. Plant and Soil 233 (2), 179–187.

Masikati, P., 2011. Improving the Water Productivity of Integrated Crop-livestock Systems in the Semi-arid Tropics of Zimbabwe: Ex-ante Analysis Using Simulation Modeling (PhD thesis). ZEF, Bonn.

Masikati, P., Manschadi, A., van Rooyen, A., Hargreaves, J., 2014. Maize–mucuna rotation: an alternative technology to improve water productivity in smallholder farming systems. Agricultural Systems 123, 62–70.

McCown, R.L., Hammer, G.L., Hargreaves, J.N.G., Holzworth, D.P., Freebairn, D.M., 1996. APSIM: A novel software system for model development, model testing, and simulation in agricultural research. Agricultural Systems 50, 255–271.

Medrano, H., Escalona, J.M., Bota, J., Gulías, J., Flexas, J., 2002. Regulation of photosynthesis of C3 plants in response to progressive drought: stomatal conductance as a reference parameter. Annals of Botany 89 (7), 895–905.

Muhr, L., 1998. The Potential of Integrating Forage Legumes into the Fallow Management of Croplivestock Farming Systems in Sub-humid West Africa (PhD thesis). Verlag Ulrich E. Grauer, Stuttgart. 146 p.

Ncube, B., Dimes, J.P., van Wijk, M.T., Twomlow, S.J., Giller, K.E., 2008. Productivity and residual benefits of grain legumes to sorghum under semi-arid conditions in southwestern Zimbabwe: unravelling the effects of water and nitrogen using a simulation model. Field Crops Research. http://dx.doi.org/10.1016/j.fcr.2008.08.001.

Nyamapfene, K., 1991. The Soils of Zimbabwe. Nehanda Publishers, Harare. 179 pp.

Powell, J.M., Pearson, R.A., Hiernaux, P.H., 2004. Review and interpretation. Crop-livestock interactions in the West African Drylands. Agronomy Journal 96, 469–483.

Reddy, A.R., Chaitanya, K.V., Vivekanandan, M., 2004. Drought-induced responses of photosynthesis and antioxidant metabolism in higher plants. Journal of Plant Physiology 161 (11), 1189–1202.

Robertson, M.J., Sakala, W., Benson, T., Shamudzarira, Z., 2005. Simulating the response of maize to previous velvet bean (Mucuna pruriens) crop and nitrogen fertilizer in Malawi. Field Crops Research 91, 91–105.

Rockström, J., Barron, J., Fox, P., 2003. Water productivity in rain-fed agriculture: Challenges and opportunities for smallholder farmers in drought-prone tropical agroecosystems. In: Kijne, J.W., Barker, R., Molden, D. (Eds.), Limits and Opportunities for Improvement, pp. 145–162.

Rufino, M., 2008. Quantifying the Contribution of Crop-livestock Integration to African Farming (PhD thesis). Wageningen UR, The Netherlands.

Rufino, M.C., Dury, J., Tittonell, P., van Wijk, M.T., Herrero, M., Zingore, S., Mapfumo, P., Giller, K.E., 2011. Competing use of organic resources village-level interactions between farm types and climate variability in a communal area of NE Zimbabwe. Agricultural Systems 104, 175–190.

Rurinda, J., van Wijk, M.T., Mapfumo, P., Descheemaeker, K., Supit, I., Giller, K.E., 2015. Climate change and maize yield in southern Africa: what can farm management do? Global Change Biology 21 (12), 4588–4601.

Schmidhuber, J., Tubiello, F.N., 2007. Global food security under climate change. Proceedings of the National Academy of Sciences 104 (50), 19703–19708.

Shamudzarira, Z., 2003. Evaluating mucuna green manure technologies in Southern Africa through crop simulation modelling (87–91). In: Waddington, S.R. (Ed.), Grain Legumes and Green Manures for Soil Fertility in Southern Africa: Taking Stock of Progress. Proceedings of a Conference Held 8–11 October 2002 at the Leopard Rock Hotel, Vumba, Zimbabwe, Soil Fert Net and CIMMYT-Zimbabwe. Zimbabwe, Harare, p. 246.

Smith, D.W., 1998. Urban food systems and the poor in developing countries. Transactions of the Institute of British Geographers 23 (2), 207–219.

Steduto, P., Hsiao, T.C., Raes, D., Fereres, E., 2009. AquaCrop-The FAO crop model to simulate yield response to water: I. Concepts and underlying principles. Agronomy Journal 101 (3), 426–437.

Touchette, B.W., Iannacone, L.R., Turner, G.E., Frank, A.R., 2007. Drought tolerance versus drought avoidance: a comparison of plant-water relations in herbaceous wetland plants subjected to water withdrawal and repletion. Wetlands 27 (3), 656–667.

Vincent, V., Thomas, R.G., 1961. An Agroecological Survey of Southern Rhodesia: Part 1 Agroecological survey. Government Printer, Salisbury.

Whitbread, A.M., Robertson, M.J., Carberry, P.S., Dimes, J.P., 2010. How farming systems simulation can aid developments of more sustainable smallholder farming systems in southern Africa. Europen journal of Agronomy 32, 51–58.

Vanlauwe, B., Coe, R., Giller, K.E., 2016. Beyond average: new approaches to understand heterogeneity and risk of technology success or failure in smallholder farming. Experimental Agriculture. http://dx.doi.org/10.1017/S0014479716000193.

Yue, B., Xue, W., Xiong, L., Yu, X., Luo, L., Cui, K., Jin, D., Xing, Y., Zhang, Q., 2006. Genetic basis of drought resistance at reproductive stage in rice: separation of drought tolerance from drought avoidance. Genetics 172 (2), 1213–1228.

Zeller, M., Diagne, A., Mataya, C., 1998. Market access by smallholder farmers in Malawi: implications for technology adoption, agricultural productivity and crop income. Agricultural Economics 19 (1), 219–229.
ZimVAC, 2013. Rural Livelihoods Assessment, Harare.

CHAPTER 14

Adaptive Livestock Production Models for Rural Livelihoods Transformation

Chrispen Murungweni[1], Obert Tada[1], Nhamo Nhamo[2]
[1]Chinhoyi University of Technology, Chinhoyi, Zimbabwe; [2]International Institute of Tropical Agriculture (IITA), Southern Africa Research and Administration Hub (SARAH) Campus, Lusaka, Zambia

Contents

14.1 INTRODUCTION

Livestock is an integral part of the principal agricultural systems in southern Africa. The future role and populations of livestock will change depending largely on climate but equally important the human population and the resource use and availability for agriculture (Alexandratos and Bruinsma, 2012; Douxchamps et al., 2016; Thornton, 2010). Climate change will affect rainfall and temperature, which in turn will affect feed production and supply of livestock products to markets. As crop–livestock systems evolve in southern Africa, suitable and innovative technologies will be required to further develop livestock systems. Similarly, policies on livestock protection, genetics and breeding, health and

Smart Technologies for Sustainable Smallholder Agriculture
ISBN 978-0-12-810521-4
http://dx.doi.org/10.1016/B978-0-12-810521-4.00014-1

nutrition will be required (Thornton, 2010). The development of climate smart technologies and policies feed directly into benefits that may accrue to smallholder farmers who rely on livestock for food and income. Climate-smart livestock production basically involves adaptation through the use of technologies that increase resilience of production and feed-base systems in a sustainable way (FAO, 2013), for example, increasing livestock productivity at minimal greenhouse gas (GHG) emission. Management of rangelands becomes important because they act as carbon sinks. The development of feed banks can be encouraged as this allows the rangeland opportunities to recover and evens out the supply of better quality feed across seasons. Improved quality of feed reduces emission of GHGs from livestock enterprises.

The greater part of planet land is suitable for livestock production. World over, livestock production seems to be decreasing except for sub-Saharan Africa (Table 14.1). Of course, figures often differ by country because of different policies, for instance, in Zimbabwe, agriculture is targeted for 12.5% growth in 2018 compared to −1.3% in 2013 with

Table 14.1 Annual livestock production growth (percent p.a.)

	1961–2007	1987–2007	1997–2007	2005/2007–30	2030–50
World	2.2	2.0	2.0	1.4	0.9
Developing countries	4.3	4.5	3.4	2.0	1.3
Idem, excl. China and India	3.4	3.6	3.5	2.1	1.5
Sub-Saharan Africa	2.5	2.8	3.3	2.7	2.6
Latin America and the Caribbean	3.2	3.8	3.8	1.6	0.9
Near east/north Africa	3.3	3.3	3.0	2.2	1.7
South Asia	3.7	3.6	3.2	2.7	2.2
East Asia	6.5	5.9	3.4	1.8	0.8
Developed countries	1.0	−0.1	0.6	0.6	0.2
Forty-four countries with over 2700 kcal/person/day in 2005/2007[a]	2.7	2.9	1.8	1.1	0.5

Aggregate livestock production was derived by weighting the four meats, milk products, and eggs at 2004/2006 international commodity prices.
[a]Accounting for 57% of the world population in 2005/2007.
Based on Alexandratos, N., Bruinsma, J., 2012. World Agriculture towards 2030/2050: The 2012 Revision, (ESA working paper).

livestock contributing more than 40% of this growth. Drought conditions and rainfall variability are set to increase in sub-Saharan Africa due to climate change, making rainfed crop production system more vulnerable relative to the more resilient livestock production system. As a result, livestock will become increasingly important in making a contribution to livelihoods of farm families. However, most of the livestock are found within the smallholder farming sector where the production objective is often risk management.

14.1.1 Mindsets Have to Shift Toward Commercialization

In smallholder farming systems, livestock are reared on grassvelds, browse, and nongrain crop remains from maize, millet, rice, and sorghum crops, and hay. Livestock contributes manure and traction power toward crop production. Livestock acts as a form of insurance against crisis times and supplies farmers with a source of regular income from sales of milk, eggs, and other products (Swanepoel et al., 2010). Therefore, the livestock production business is conducted for a range of values: as a form of wealth, prestige, or business.

Livestock should contribute more to people's livelihoods and nations' gross domestic product. Farming for subsistence and risk management have several disadvantages. Livestock numbers become difficult to control if risk management remains the main objective. With the climate changing toward drier conditions, greater proportions of livestock will be held for risk cover. Also, people tend to adjust for loss in crop production by investing more in livestock. Large numbers would present some sense of security and pride. However, increasing livestock numbers in the face of climate change oriented toward reduced feed base can lead to poor livestock productivity, lower meat yield and quality, poor health, increased vulnerability of rangelands to uncontrolled stocking rates, and reduced livelihoods due to reduced feed-base, increased disease outbreaks, and increased livestock deaths. Intensification (Fig. 14.1) requires commercial feeds that can be out of reach of most smallholder farmers because of high costs. This scenario promotes the mindset of keeping livestock for risk aversion and will forever keep off-take from the smallholder farming sector at its lowest. Sub-Saharan Africa cannot develop as anticipated when the major resource is not being mobilized to generate income and employment. The major question therefore is, "How can livestock in smallholder sector be mobilized for local households to generate income to diversify of their livelihoods?" What is needed for people to commercialize their livestock? Who should commercialize what?

Figure 14.1 Pictures of livestock farms in southeastern Zimbabwe where both improved breeds of cattle and goats are produced. *Photos by Chrispen Murungweni.*

Livestock plays key sociocultural roles within rural communities. Livestock development projects targeted for improving rural livelihoods through commercialization should take sociocultural roles of livestock into consideration. In addition to complex ownership patterns, low productivity is a major challenge for the majority of livestock households. Uncoordinated planning for marketing results in low off-take, poor marketing networks, under developed marketing infrastructure, and poor access to livestock markets. Such conditions are detrimental to the development of the livestock industry; the starting point should be giving confidence to the farmer that their breeding stock is safe under different climatic conditions for them to feel secure even when selling periodically. This is possible if the famers are knowledgeable about advantages of increasing productivity over increasing numbers, if they can understand importance of self-organization and consistency and persistence of supply line of quality product over crisis disposal of poor quality materials.

14.2 LIVESTOCK MANAGEMENT IN A CLIMATE-SMART AGRICULTURAL ENVIRONMENT

The strategies used by farmer in livestock production largely depend on the resources available and the level of development of the livestock value chains. Climate plays a crucial role in the definition of the livestock systems applicable in each location. Table 14.2 shows some of the systems that are applicable across a range of environments.

Climate change can be expected to have several impacts on feed crops and grazing systems, including the following (Hopkins and Del Prado, 2007):

Table 14.2 Livestock systems according to the classification of Seré and Steinfeld (1996)

Generic	Specific	System
LG (livestock only)	LGA	Livestock only systems, arid–semiarid (LGP <180 days)
	LGH	Livestock only systems, humid–subhumid (LGP <180 days)
	LGT	Livestock only systems, highland/temperate[a]
MR (mixed rainfed)	MRA	Mixed rainfed crop/livestock systems, arid–semiarid (LGP <180 days)
	MRH	Mixed rainfed crop/livestock systems, humid–subhumid (LGP <180 days)
	MRT	Mixed rainfed crop/livestock systems, highland/temperate[a]
MI (mixed irrigated)	MIA	Mixed irrigated crop/livestock systems, arid–semiarid (LGP <180 days)
	MIH	Mixed irrigated crop/livestock systems, humid–subhumid (LGP <180 days)
	MIT	Mixed irrigated crop/livestock systems, highland/temperate[a]
LL (landless)	LLM	Landless monogastric systems
	LLR	Landless ruminant systems

[a]Temperate regions: areas with one or more months with monthly mean temperature, corrected to sea level, of less than 5°C. Tropical highlands: areas with a daily mean temperature, during the growing period of 5–20°C.

- Changes in herbage growth brought about by changes in atmospheric CO_2 concentrations and temperature;
- Changes in the composition of pastures, such as changes in the ratio of grasses to legumes;
- Changes in herbage quality, with changing concentrations of water-soluble carbohydrates and N at given dry matter (DM) yields;
- Greater incidences of drought, which may offset any DM yield increases;
- Greater intensity of rainfall, which may increase N leaching in certain systems.

14.2.1 Livestock–Crop Interactions

In southern Africa crop–livestock farming systems are dominant over other farming systems (Mcintire et al., 2016). Under this system, land use is a characterized area under annual and perennial crops and area reserved for livestock. Compared to the large-scale sector, the allocation of land resources is not demarcated such that during off-season for crops livestock graze crops remain on the fields. Livestock manure and provision of draught power are the major linkages between the crops and the livestock sections of agricultural production. Manure, in particular, is a key resource that links the nutrient cycles on the grass veld and on the cropping fields (Herrero et al., 2015; Mcintire et al., 2016). Yet it is widely known that GHG emissions from manure constitute about 40% of methane, a significant amount of the total emissions from agriculture systems (Herrero et al., 2013). Climate extremes threaten the productivity of both cropping and grassland on the one hand, whereas livestock manure production and management exacerbate the occurrence of extremes. Sustainable crop livestock systems require both mitigation and adaptation practices to maintain production for an increasing population.

14.2.2 Manure as a Key Resource in Crop–Livestock Systems

Manure use as a soil fertility input for smallholder farmers has been documented by numerous researchers (Mugwira and Murwira, 1997; Murwira et al., 1995; Nhamo et al., 2003; Nyamangara et al., 2003). The overall effect of manure on crop production has been explained by the intrinsic quality characteristic of the manure as influenced by the livestock systems, management, curing, and storage (Nhamo et al., 2003). Major benefits from manure have been as a result of the contribution to the nitrogen (N) economy on cropping systems (Mugwira and Mukurumbira, 1984; Nyamangara et al., 2003). High-quality manure from dairy feedlots has shown the potential use in a substitutive manner with mineral fertilizer sources (Nhamo, 2011). Fig. 14.2 shows that when high-quality manure is used in combination with mineral fertilizer the overall fertilizer-use efficiency was relatively comparable across the combination applied on maize. The use of manure needs to be carefully reviewed in view of its contribution to methane production and potential environmental pollution, especially where large herds of livestock are involved.

14.2.3 Manure Emissions and Pollution

Livestock produces significant amounts of methane, a GHG that contributes to climate variability. Enteric fermentation and animal wastes contribute about 80 and 25 million tons respectively of methane (total 105 million tons) annually, making them the major source of emission from agriculture (Moss et al., 2000). Several factors determine the amount of CH_4 production in the livestock systems. Factors such as level of feed intake, type of carbohydrates in the diet, feed processing, and alterations in the ruminal

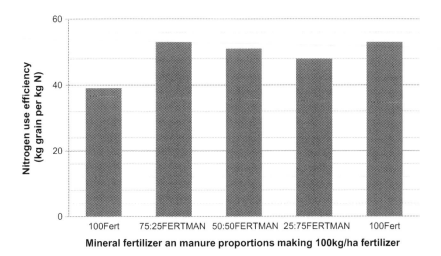

Figure 14.2 Nitrogen use efficiencies of high-quality dairy feedlot cattle manure and mineral fertilizers on maize in Zimbabwe.

microflora (Johnson and Johnson, 1995). Manipulation of these factors can reduce methane emissions from cattle. Slurries are also a significant source of NH_3 and N_2O emissions to the atmosphere (Amon et al., 2006). Dominant emissions of methane occur during storage, whereas nitrous oxide emissions largely occur during field application of manures. Several methods have been put forward to reduce and/mitigate GHG emissions, which include the use of organic material that acts as absorbent in manure slurries.

High-quality manure when not disposed of properly can cause pollution. The high N and P content in the manure either leaches into soil profiles or washes away into water bodies where they cause eutrophication. Therefore, there is a need for proper management of feedlot heaps of manure besides the GHG emissions.

14.2.3.1 Management of (In)breeding in Livestock Production Systems

Breeding is a very important driving force in adaptation to future production conditions. Breeding has made large contributions to improved efficiency and thus indirectly progressing to smart climate agriculture. None the less, there is little evidence that animal breeding organizations have considered climate change adaptation in their definition of breeding objectives and breeding schemes. Inbreeding is technically defined as the mating of animals more closely related than the average relationship within the breed or population concerned. It is the increase in the probability that an offspring receives the same gene from both parents. In general, inbreeding results in an overall lowering in performance. It is most obviously reflected in poorer reproductive efficiency, including higher mortality rates, lower growth rates, and a higher frequency of hereditary defects. Despite these generally harmful effects, inbreeding is a very useful tool in the field of animal agriculture. It enables the breeder to uncover and eliminate harmful recessive genes within the population, thus making the livestock more adaptable to the prevailing climatic conditions. It is also essential to the development of prepotent animals and is desirable in the development of distinct family lines. In addition, seed stock and commercial producers have successfully used line breeding to maintain a degree of genetic relationship in their animals to some outstanding ancestor or ancestors.

The decrease in fitness that results from such inbreeding is known as *inbreeding depression*. Inbreeding depression is thought to be caused primarily by the collection of a multitude of deleterious mutations, few of them fatal, but all with diminishing fitness. Normally, in an out-breeding population these alleles would be selected against, hidden, or corrected by the presence of good alleles in the population. This explains why we do not all have many genetic diseases as sexual reproduction and shuffling of alleles of genes occurs when two unrelated individuals mate. Inbreeding depression encompasses a wide variety of physical and health defects that warrants careful management in achieving smart climate agriculture. Any given inbred animal generally has many defects including elevated incidence of recessive genetic diseases, reduced fertility both in litter size and in sperm viability, increased congenital defects such as cryptorchidism, heart defects and cleft palates, fluctuating asymmetry (such as crooked faces, uneven eye placement and size), lower birth weight, higher neonatal mortality, slower growth rate, smaller adult size, and loss of immune system function.

Livestock production contributes to climate change through GHG emissions especially of methane gas by ruminants. GHGs are often blamed for increasing environmental temperatures. Livestock is, however, affected by climate change through the physiological effects of higher temperatures on individual animals; as a result, geographically restricted rare breed populations will be badly affected. Indirect effects may be felt via ecosystem changes that alter the distribution of animal diseases or affect the supply of feed. In sub-Saharan Africa where climate change models predict increased temperatures due to climate change, breeding goals need to be adjusted to account for higher temperatures, lower quality diets, and greater disease challenges. Species and breeds that are well adapted to such conditions may become more widely used. Climate change mitigation strategies, in combination with ever increasing demand for food, may also have an impact on breed and species utilization, driving a shift toward monogastrics and breeds that are efficient converters of feed into meat, milk, and eggs. This may lead to the neglect of the adaptation potential of local breeds in developing countries. Given the potential for significant future changes in production conditions and in the objectives of livestock production, it is essential that the value provided by animal genetic diversity is secured. This calls for better characterization of breeds, production environments, and associated knowledge including the compilation of more complete breed inventories, improved mechanisms to monitor and respond to threats to genetic diversity, more effective in situ and ex situ conservation measures, genetic improvement programs targeting adaptive traits in high-output, and performance traits in locally adapted breeds, increased support for developing countries in their management of animal genetic resources, and wider access to genetic resources and associated knowledge.

A practical approach for minimizing inbreeding and maximizing genetic gain is practiced in dairy cattle where the best bull is selected for each cow. This is based on the offspring estimated breeding values minus inbreeding (F) depression to reduce the average inbreeding of the calves by about a third to half even when the cost of inbreeding depression is as low as 3%.

In controlling inbreeding in modern livestock breeding programs, accurate breeding value estimation, and advanced reproductive technologies are of paramount importance. Such programs lead to rapid genetic progress and accumulation of inbreeding through more impact on a few selected individuals or families. Inbreeding rates are accelerating in most species, and economic losses due to inbreeding depression in production, growth, health, and fertility are a serious concern. Most research has focused on preservation of rare breeds or maintenance of genetic diversity within closed nucleus breeding schemes. However, the apparently large population size of many livestock breeds is misleading because inbreeding is primarily a function of selection intensity. Corrective mating programs are widely used in some species, and these can be modified to consider selection for economic merit adjusted for inbreeding depression. Selection of parents of artificial insemination bulls based on optimal genetic contributions to future generations, which are a function of estimated breeding values and genetic relationships between selected individuals, appears most promising. Rapid implementation of such procedures is necessary to avoid further reductions in effective population size.

Fortunately, most breeders are well aware of the negative effects of severe inbreeding, and few breeders would purposefully breed siblings with one another. However, another strategy, sometimes called line breeding, is very common, and is practiced. Line breeding involves mating within a historically defined and related set of individuals (a line). Lines may be formally or informally recognized and there may be lines within lines. Within these family lines, some people may maintain or promote even smaller lines. These lines are maintained because each line has (or is perceived to have) unique and desirable traits, e.g., disease resistance. However, the degree of inbreeding is related not just to the immediate relationship between the two mated individuals but also to the degree to which these individuals were already inbred. For instance, if a particular bull is widely used in a line-breeding program, then many of the cattle in that program will already share the genes of that bull and the chances of getting two deleterious alleles from that bull in the same offspring are greatly increased. Even small increases in inbreeding result in some inbreeding depression. For instance, just a 1% increase in inbreeding results in a measurable decrease in milk quantity and quality, shortening of productive life, and increase in calving interval in studied breeds of cattle.

The effects of inbreeding can also be avoided to achieve climate smart agriculture. When two lines are crossed, then any deleterious alleles present in one line but not in the other are masked, and there is typically a boost in the fitness of the offspring. This effect is known as heterosis or hybrid vigor and is essentially the opposite of inbreeding. This effect even extends to crosses between breeds of cattle, which is why cross-breeding programs are popular. Reproductive technologies currently available allow high selection intensity in most livestock species and are combined with selection methods, which take into account family information, such as best linear unbiased prediction. As a result, response to selection has been enhanced but rates of inbreeding have also increased and they currently represent a serious concern for several breeding programs due to the possible consequences in terms of inbreeding depression and genetic variability. Therefore, methods have been proposed where selection is carried out by appropriately weighing the predicted breeding value of an individual and the inbreeding generated in the population. Pure breeders and line breeders use inbreeding to select for a trait, which they seek to promote. As the relatedness (homozygosity) increases due to inbreeding some deleterious recessive traits begin to manifest in your herd, for example, inbreeding severely reduces milk yields over time as the homozygosity of the livestock increases. In beef production, important production traits such as conformation, weaning weights, weaning age, age at first calving, and postweaning gains are all affected by inbreeding.

14.2.3.2 Development of Livestock Markets Under Fragmented Livestock Marketing Environments

The potential of the largest proportion of farmers to commercialize livestock lies on small stock. Almost every household in sub-Saharan Africa owns some form of poultry, a bigger proportion keep small stock (to include small ruminants) and cattle. It is difficult to involve farmers in formal marketing if the farmers are not organized enough to develop sustainable marketing channels. Farmers keep different types of livestock for various diverse reasons: prestige, risk aversion, drought power, food, butter exchange, etc. The numbers are also diverse per individual household. Making a typology of households is the first most important start to develop a viable marketing value chain. Typology facilitates the correct targeting with interventions for vibrant innovation within communities. If farmers organize themselves and their system in a way as proposed in Fig. 14.3 below, commercialization of livestock can be possible. This model seeks to develop an organized breeding and conservation, product processing, and marketing system for livestock by farmers themselves.

Typical of the economies of most developing countries, the noncommercial smallholder farmers use an informal or spot on marketing system, which is more often than not, a system of necessity than a system of choice. Informal markets hold potentially many risks to both the producer and the consumer, whereas commercializing and formalizing the production and marketing of livestock in the smallholder areas will require a vast outlay of capital resources, but gains on infrastructure, institutions, legal frameworks, markets, novelty in product development, capacity building, and technology transfer will empower these people. The proposed model (Fig. 14.3) aims at promoting a shift from subsistence activities that are commonly a characteristic of rural households, especially the poor, to a focus on a business venture that has the potential to generate enough income to improve rural livelihoods.

Most rural farmers keep small herd sizes of livestock ranging from 1 to 20 in large stock and one to over a 100 for small stock. Small numbers result in high transaction costs for the smallholder livestock producers to enter the commercial market. The proposed business venture (Fig. 14.3) has a special characteristic of pulling resources together by participating households. It encourages networking within farmer groups and between farmers and

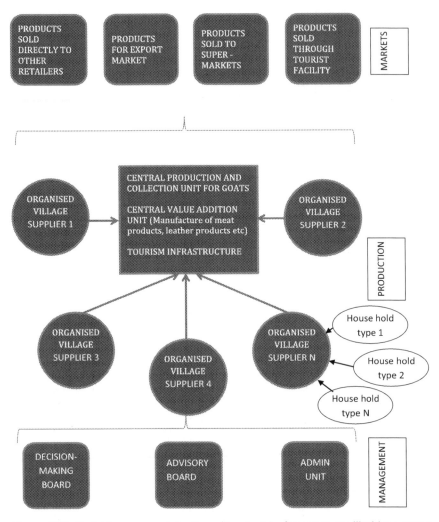

Figure 14.3 Model for commercialization of livestock by farmers in smallholder systems of sub-Saharan Africa.

their stakeholders. It strengthens linkages at various levels, for example, households on low performing systems will learn from those on high performing ones during field days and demonstrations. Research will provide the vehicle of shift from a low performing system to a high performing one. Stakeholder forums will play a key role for timely feedback and provides the platform for specification of roles for stakeholders.

The proposed setup will support the development of the livestock value chains on production, and products processing and set appropriate strategies, regulations, and institutional arrangements necessary for their commercialization. The issue of breed conservation has attracted attention at a global level, whereby the responsibility is bestowed on the country of origin of the breed. The indigenous breeds may meet the future market demands and conservation shall be both in situ and ex situ. Marketing of processed products of livestock will help farmers by creating more reliable marketing chain. The set-up of the proposed livestock marketing system will enhance the delivery of livestock inputs and services to livestock farmers through the setting-up of the livestock breeding and conservation units, targeted extension services, research, and better linkages between stakeholders. The establishment of localized livestock processing units at the central point provides an important marketing infrastructure giving reason for livestock households to shift their mindsets from subsistence to commercialization. Several strategically organized trainings will help in strengthening the capacity of goat farming communities and their link with private sector. The establishment of stakeholder forums will help strengthen the communication between private and public institutions to provide better services to the livestock sector.

14.2.4 Research Gaps

Some of the knowledge gaps of climate change impacts on livestock-based systems and livelihoods of livestock farmers exist and these need urgent attention for the development of livestock value chains:

- Rangelands and feed quantity and quality: The impact of climate extremes will affect the primary productivity, veld species distribution, and can change the overall nutrition for livestock. As species either survives or becomes extinct due to CO_2 and other reasons the feed quality and quantity will remain important questions for livestock planners. Species composition will be shaped by competitive factors leading to changes in the carrying capacities of rangelands (Thornton et al., 2009).
- Potential for mixed crop–livestock systems: Productivity in both the subsystems will affect the overall outlook of agriculture. Climate impacts that are localized will have an effect on primary productivity; there could be shifts in harvest indexes and the availability of stover and crop residues for livestock will change thereby affecting the production.
- Heat stress: Temperature extremes will become more frequent and hence the average temperatures may increase in some locations. It is not clear to what extent extreme heat will affect livestock production?

- Diseases and disease vectors: Climate change will affect the prevalence of pest, vectors, and diseases. Livestock movements based on the disease challenges will need to be restricted further. This will affect breeding programs and more strict regulations will need to be enforced and this may affect the development of livestock.
- Biodiversity will respond to climate extremes and this may challenge the current benefits from ecological biodiversity; however, the change in species densities as systems change is not known.
- Projecting future needs: Animal breed biodiversity is based on the current germplasm, will genetic resources will be useful, superior, and adaptive?
- Ecosystem services associated with livestock systems: Climate change may enhance services from livestock and how these changes on ecosystems goods and services will occur are not clear.
- Indirect impacts: Human beings consume livestock products and in southern Africa this is related to increase in disposable income. How will human health impacts of climate change intertwine with livelihood systems and vulnerability?

14.3 CONCLUSION

Livestock production is affected by population and availability of land for grazing. Climate change will affect the quality of feed produced in smallholder farming systems. Adaptation and mitigation of GHG are required to improve the efficiency of livestock systems. Breeding is required to preserve the gains of yesteryears and to build on the successes to produce and support climate smart livestock systems.

REFERENCES

Alexandratos, N., Bruinsma, J., 2012. World Agriculture towards 2030/2050: The 2012 Revision. (ESA working paper).

Amon, B., Kryvoruchko, V., Amon, T., Zechmeister-Boltenstern, S., 2006. Methane, nitrous oxide and ammonia emissions during storage and after application of dairy cattle slurry and influence of slurry treatment. Agriculture, Ecosystems & Environment 112, 153–162.

Douxchamps, S., Van Wijk, M.T., Silvestri, S., Moussa, A.S., Quiros, C., Ndour, N.Y.B., Buah, S., Somé, L., Herrero, M., Kristjanson, P., 2016. Linking agricultural adaptation strategies, food security and vulnerability: evidence from West Africa. Regional Environmental Change 16, 1305–1317.

FAO, 2013. "Module 8: Climate-Smart Livestock" in Climate-Smart Agriculture Sourcebook. (Rome, Italy).

Herrero, M., Havlík, P., Valin, H., Notenbaert, A., Rufino, M.C., Thornton, P.K., Blümmel, M., Weiss, F., Grace, D., Obersteiner, M., 2013. Biomass use, production, feed efficiencies, and greenhouse gas emissions from global livestock systems. Proceedings of the National Academy of Sciences 110, 20888–20893.

Herrero, M., Wirsenius, S., Henderson, B., Rigolot, C., Thornton, P., Havlík, P., De Boer, I., Gerber, P.J., 2015. Livestock and the environment: what have we learned in the past decade? Annual Review of Environment and Resources 40, 177–202.

Hopkins, A., Del Prado, A., 2007. Implications of climate change for grassland in Europe: impacts, adaptations and mitigation options: a review. Grass and Forage Science 62, 118–126.

Johnson, K.A., Johnson, D.E., 1995. Methane emissions from cattle. Journal of Animal Science 73, 2483–2492.

Mcintire, J., Bourzat, D., Prabhu, P., 2016. Crop-Livestock Interaction in Sub-Saharan Africa. World Bank, Washington, DC.

Moss, A.R., Jouany, J.P., Newbold, J., May 2000. Methane production by ruminants: its contribution to global warming. Annales de Zootechnie 49 (3), 231–253.

Mugwira, L., Mukurumbira, L., 1984. Comparative effectiveness of manures from the communal areas and commercial feedlots as plant nutrient sources. Zimbabwe Agricultural Journal (Zimbabwe) 816, 241–250.

Mugwira, L., Murwira, H., 1997. Use of Cattle Manure to Improve Soil Fertility in Zimbabwe: Past and Current Research and Future Research Needs. Centro Internacional de Mejoramiento de Maiz y Trigo (CIMMYT), Mexico DF, Mexico.

Murwira, K., Swift, M., Frost, P., 1995. Manure as a key resource in sustainable agriculture. In: International Conference on Livestock and Sustainable Nutrient Cycling in Mixed Farming Systems of Sub-Saharan Africa, Addis Ababa (Ethiopia), November 22–26, 1993. ILCA.

Nhamo, N., Mupangwa, W., Siziba, S., Gatsi, T., Chikazunga, D., 2003. The Role of Cowpea (*Vigna unguiculata*) and Other Grain Legumes in the Management of Soil Fertility in the Smallholder Farming Sector of Zimbabwe. Grain Legumes and Green Manures for Soil Fertility in Southern Africa: Taking Stock of Progress, pp. 119–127.

Nhamo, G., 2011. REDD+ and the global climate policy negotiating regimes: challenges and opportunities for Africa. South African Journal of International Affairs 18, 385–406.

Nyamangara, J., Piha, M., Giller, K., 2003. Effect of combined cattle manure and mineral nitrogen on maize N uptake and grain yield. African Crop Science Journal 11, 289–300.

Seré, C., Steinfeld, H., 1996. World Livestock Production Systems-Current Status. Issues and Trends. FAO, Rome, p. 83.

Swanepoel, F., Stroebel, A., Moyo, S., 2010. The Role of Livestock in Developing Communities: Enhancing Multifunctionality. AFRICAN SUN MeDIA.

Thornton, P., Van De Steeg, J., Notenbaert, A., Herrero, M., 2009. The impacts of climate change on livestock and livestock systems in developing countries: a review of what we know and what we need to know. Agricultural Systems 101, 113–127.

Thornton, P.K., 2010. Livestock production: recent trends, future prospects. Philosophical Transactions of the Royal Society of London B: Biological Sciences 365, 2853–2867.

CHAPTER 15

Delivering Integrated Climate-Smart Agricultural Technologies for Wider Utilization in Southern Africa

Nhamo Nhamo, David Chikoye, Therese Gondwe
International Institute of Tropical Agriculture (IITA), Southern Africa Research and Administration Hub (SARAH) Campus, Lusaka, Zambia

Contents

15.1 INTRODUCTION

Agricultural improvement has increased chances of human advancement on one hand, whereas on the other hand, certain practices may threaten the environmental quality and hence sustainability. This is viewed as the dilemma of agro-ecosystems managers. Therefore, striking a balance between the aspirations of humanity and systems sustainability places a huge responsibility on stakeholders in agro-ecosystems across the whole spectrum from the plot, catchment, continental, and global levels. Environmental protection, ensuring production potential of systems for the future while meeting the current demand for food for health and nutrition, incomes to reverse poverty, raw materials for profitable industries will remain vitally important going forward. Access to knowledge of improved smart technologies through adequate extension and training of least developed smallholder

Smart Technologies for Sustainable Smallholder Agriculture
ISBN 978-0-12-810521-4
http://dx.doi.org/10.1016/B978-0-12-810521-4.00015-3

farmers can save the agro-ecosystems from degradation. Farmers, especially youth and women, in southern Africa stand to benefit from the investment into the application of candid smart agricultural technologies to daily farm operations.

A projection of the world population of 9 billion inhabitants by 2050 (Godfray et al., 2010) challenges the current production systems and doubling the production has been suggested as the feasible solution to feeding the growing population. The major challenge is that in southern Africa the population growth rates exceed the agricultural growth rate (Tilman et al., 2002). To date climate extremes impose a strong barrier for increasing yields for smallholder farmers in drought-prone areas with constant extreme temperature incidences. Besides climate variability and change, rural–urban, cross-country, and cross-continent migration for various reasons also threaten labor in terms of its availability and cost. There is a need to analyze food delivery systems, in view of immigration trends, the population of the hungry poor farmers, income for farmers, and support jobs in agro-industries. How to manage the population fluxes across different countries in the region within the limits of climate change is not clear?

Doubling food production is a tall order for most countries, agricultural intensification has been suggested as the viable option. The pertinent question, which requires our attention now and in the future, is how sustainable crop and livestock intensification can be achieved given the climate variabilities and change. Without claiming an exhaustive menu of priority areas, delivering smart technologies will inevitably require investment in the following:

1. *Increased human resources capacity for delivery:* The capacity building on sectors supportive of agricultural development will be key if a knowledge-centered approach should be successfully implemented for agriculture transformation. The demographics of countries in southern Africa has a young population. Therefore, sustained and strategic investment in the development of an entrepreneurial work force in agriculture increases the probability of reaping the demographic dividend benefits. Human capacity improvement will enhance the quality of service delivery, resource-use efficiency, and appropriate targeting where necessary. Recent commitments by governments to resolve climate variability impacts need to be channeled into building human resource capacity in key sectors including agriculture to develop more options for adaptation and mitigation of climate change.

2. *Investment base on the per capita needs in southern Africa*: The future food requirements have been projected to increase for the majority of African countries and how governments will organize communities in response to the higher demands is an essential question. Sustainable food production, processing systems, and consumption patterns need to be informed by the population growth figures for the subregion. Other than food, power, housing, communication technologies, mechanization, and infrastructure development are realistic needs required for societal development. Smart technologies have a good chance of being utilized in some of these investments to address the future needs of individuals and communities.

3. *Smart technologies based agricultural systems*: Modern technologies, based on empirical research findings, which are climate extreme resilient, are needed for agriculture to deliver jobs, food, and incomes, which all improve the livelihoods of poor farmers. Climate resilience has to be supported by improved germplasm of crop and breeds of livestock, which can withstand the heat waves, extreme cold weather, drought and floods, and high CO_2 emissions in the atmosphere (Chapter 4). Second, the intensification of crop–livestock systems has to be supported by improved management practices that target increased input efficiencies, e.g., fertilizer, water, and agronomic-use efficiencies. Depending on the scale of operations, farmers need to access these technologies without delay so that the anticipated delivery can occur at both small and large scales. Similarly, processing and value addition are areas in which more investments are required. Better methods of processing and adding value increase utilization and incomes from the farm produce compared to strictly the supply of raw materials (Chapter 10).

15.2 LINKING SMART TECHNOLOGIES

Technologies for agricultural transformation were developed in the form of silos and the application of smart technologies has to be systematic, integrated, and sustainable. By design rather than by default efforts are required to bring together the suite of practices to meet the demand for food and to reduce environmental and financial externalities. Climate variability and change will require multitechnology, multistakeholder, and concerted efforts to overcome the challenges it poses on the agricultural production. Several work packages need to be revisited as research and development experts seek solutions to climate change extremes.

Work package on cultivars and breeds: Breeding pipelines for climate smart crop cultivars and livestock breeds.

Crop cultivars that have the capacity to escape and/or tolerate drought are currently sparingly used by smallholder farmers. Several reasons to explain this pattern, including limited access or absence of such cultivars on the market, lack of fit to the growing season length, high cost for resource-poor farmers little overlapping with local taste preferences, and erratic supply of seed across years (Marra et al., 2003; Nkonya et al., 1997). There is, therefore, a need to scale up and use and develop new cultivars to resolve the challenges faced by farmers as a result of climate impacts. Breeding pipelines supporting the development of diverse crops for humid zones (mostly highlands) and dry areas need to be strengthened (Araus and Cairns, 2014). Breeding for tolerance to extreme events and positive response to management practices could be difficult; however, marker-assisted selection for traits may lead to improved results. Pyramiding of stable quantitative trait locuses controlling component traits may contribute to solutions (Witcombe et al., 2008). New genomic technologies hold promise to new discoveries leading to improvement. Molecular markers can be used to identify quantitative trait locis (QTLs) in maker-assisted selection using modern methodologies, which seek to evaluate QTLs, which cosegregate with yield components under extreme climate (Araus and Cairns, 2014; Witcomb et al., 2008). Currently maize receives a lion's share of resources from the public and private sector, whereas limited resources are allocated for small grains, legumes, and root and tubers crops. Both legumes and root crops have several advantages, including multiple uses for food and industrial application. These advantages need to be harnessed so that smallholder farmers can benefit.

Similarly, crops are prone to pests: superabundance of white flies and green mites in cassava; aphids and pod sackers in cowpea; stalk borers in maize, which require a combination of improved cultivars and good crop husbandry as solutions to these biotic challenges. Diseases such as cassava mosaic disease and cassava brown streak disease both transmitted by the white flies; leaf blight and cowpea mosaic in cowpea; rust and frog eye in soybean; cob and stalk rot in maize will require urgent attention to reduce climate change-driven biotic stress on crops (Chakraborty et al., 2000; Jarvis et al., 2012; Karungi et al., 2000; Nichols, 1950; Rosenzweig et al., 2001). A combination of technologies is required to avert the projected impact of biotic constraints to production and processing of food crops (Schmidhuber and Tubiello, 2007).

Better performing breed in livestock systems will be required to sustain production under extremes of weather and climate (Thornton et al., 2009). Smallholder farmers in dry areas store value in small livestock, e.g., goats while large livestock is important for large-scale cattle farmers. In some countries livestock is a source of prestige and cultural value. Appropriate breeds are also required for a range of farmers who practice mixed crop–livestock farming, and they operate in a different ecosystem. Feed efficient and disease-tolerant herds will assist farmers in managing the stocking rates in climate-affected zones.

Major gains have been reported in the area of livestock disease control (Perry et al., 2013); however, climate change will affect the incidence of pests and vectors (Thornton et al., 2009) and hence more needs to be done. Increased effectiveness of livestock programs in Botswana and Namibia can increase the productivity of livestock in the region (Hitchcock, 2002).

Work package on systems agronomy and management: Increasing input use on soils and cropping systems for crop intensification.

Droughts caused by climate variability and changes will reduce the efficiency with which soil systems support crop production and the manner in which input resources are utilized in cropping systems. Climate smart cropping systems are needed and improved management of soil systems is required for successful crop intensification in changing climates. Both organic and mineral fertilizer sources need to be evaluated for increased efficiency and synchrony between nutrients and moisture regimes in soils. Similarly, the use of legumes in boosting soil N budget through biological nitrogen fixation need to be managed in relation to the use of P, inoculants (e.g., Nhamo et al., 2014), and other soil nutrient elements to mitigate climate change through the reduction of emissions from the crop lands.

Management of soil organic matter under sustainable crop intensification will contribute to carbon sequestration (Lal, 2004), leading to the reclamation of degraded soil and maintenance of productive fields. Both surface management and incorporation of organic materials into the soil have been identified as important methods of improving soil organic matter. Categories of organic materials available on the farm range from crop residues to composted manures all are important inputs to soil management practices (Paustian et al., 2000). Investing in soil organic matter will lead to sustainable soil use in the long run.

Crop intensification will rely on cultivars, which respond to the management of soil nutrient and water under different environmental conditions. Response to the fertilizer application of NPK, organic manures, pH regulation, and use of rhizobia inoculation and mycorrhizal associations is very important in halting nutrient mining and enhancing soil health. Equally important is the response of cultivars to micronutrients and other agro-chemicals including lime, pests, disease, and weed control methods (Batáry et al., 2012).

Improving the management of production systems will provide the main frame through which the effect of agricultural intensification will lead to secure food sources for the majority of people. However, a combination of limited resources and challenging climate will necessitate the development of methods to reduce waste as most of the potential food goes to waste in communities without infrastructure for food storage and preservation. The proportion of food wasted from farm gate throughout the whole value chain at present accounts for more than half of the produce. Food policies that favor reduction in losses can contribute to eradication of hunger and better distribution and pricing of food stuffs (Hanjra and Qureshi, 2010). Similarly overreliance on one crop value chain and following close narrow diets either as a matter of preference or as a result of tradition can lead to more hunger and poverty. Shifting diets to more efficiently produced foods and climate resilient crops will be a topic for future debates. In southern Africa, maize dominates food systems and the demonstrated inability to withstand droughts and floods is a major cause for concern.

Alternative and competing uses of major crops such as maize, sugarcane, or cassava also threaten fragile farm communities with hunger. The use of staple crops for biofuels has increased in recent years (Naylor et al., 2007). The benefits of biofuels in mitigation of climate change, alleviation of fuel shortages globally, and its support for rural development have been the major drivers for this sector (Fischer et al., 2009; Tilman et al., 2009). However, the research has shown that the recent boom in biofuels as a result of collaborative efforts between the energy and agriculture sectors may lead to higher prices and limited access to food, thereby affecting livelihoods of the poor (Mitchell, 2008; Ajanovic, 2011).

Work-package climate forecasting and risk reduction: Increased access and use of weather forecasting information for developing smart responses to reduce the risk of climate change.

The quality of weather forecasting information that farmers receive is very critical in developing a response for impending bad weather, which can destroy crops. Currently, there is very little interface between the widely shared weather forecasting information in the public domain and responses by farmers in terms of informed crop management strategies; hence, the need for improvement. The weather information from the meteorological services departments needs to assist farmers to develop the schedules and respond timely and adequately to climate risk (Tol et al., 2003). Coupled with the indigenous knowledge systems on weather changes (Nakashima et al., 2012; Nyong et al., 2007), forecasting can provide useful guides that will support crop intensification models. Collaborative linkages are required among crop specialists working on research for development, meteorological services departments, and farmers with the aim of sharing the most useful pieces of information that can be related to reducing the risk of crop failure as a result of extreme weather incidences.

Crop simulation modeling can also be used in designing cropping calendars for farmers in areas where weather patterns repeatedly affect crops. A combination of biophysical and economic models can be used for technology and farmer typology targeting to improve effectiveness of scaling out and the technologies themselves (Semenov and Porter, 1995). These models, however, are dependent on the available information on weather, soils, socioeconomic data of farmers, and the performance of the technologies. Risk from modeling exercises can then be calculated on the soil and environment, inputs including land and agrochemicals.

Work-package on human performance improvement: Capacity building.

To adequately learn from experiments and promotion of smart climate technologies, there is a need for capacity building of both government agents' staffs and those from the private sector on critical issues and practices. There are possibilities that the framework of analysis of technologies will change to address directly the challenges that agriculture will face under climate variations. There will be need for capacity development activities at several levels to support.

15.2.1 Short Courses

Climate smart agriculture courses will be required for upgrading skills for technicians in areas such as breeding, agronomy, integrated crop management systems, geographical information systems applications, seeds production, and integrated pests management systems. Particularly youth, women, and some male farmers in southern Africa will be included as they are often the least informed when it comes to modern farming methods. The targeting of youth and women as special groups of farmers who need to be supported highlights the important contribution, which these groups can make by utilizing climate smart technologies in their farming operations. Work with these groups will focus specifically on the development of climate smart agribusiness enterprises, which can directly impact household incomes in a sustainable manner.

15.2.2 Postgraduate Degree Training

Regional university programs on agro-climatology, physics, and climate studies are available in major universities in Malawi, Namibia, South Africa, Zambia, and Zimbabwe. The impact of these programs in generating the critical mass required to tackle climate change and variability challenges has been challenged. Until recently only a few scientists at M.Sc. and Ph.D. level have been working directly in climate issues as they are related to agricultural development. There is, therefore, an apparent need for the region to shift attention and support the development of human capacity in the area of climate smart agriculture innovations. To strengthen analytical skills further, stronger collaborative linkages with advanced research institution can enhance the process.

Work package developing working partnerships and collaborations: Linkages developed and strengthened among different partners to accelerate transfer and adoption of climate smart technologies.

For the sustainability of climate smart agriculture outcomes, there will be a need for improving collaboration among countries and development partners, research for development, farmers and extension workers on climate smart technologies. Employment of the innovation systems approach can enhance partnerships while at the same time linking farmers to market for the promoted crops. Innovation platforms will be formed, where they are not formed already to support the development of the value chains of the promoted crops. To speed up the adoption of technologies, a range of participatory extension approaches will be applied while working with partners from stakeholders and using the lead farmers approach (Fig. 15.1).

15.3 RETHINKING ORGANIZING VALUE CHAIN ACTORS FOR EFFICIENT SYSTEMS

Climate variation will reduce the periods within a cropping season when favorable conditions will prevail and this requires that the cropping and livestock systems be more efficient in converting available resources to biomass and eventually yield. Similarly, value chains need to be more efficient leading to the reduction in food wastages. More efficient value chains will increase benefits to farmers who lose most of their produce on the market floors. The overall price of some commodities is driven by the losses incurred on uninsured goods that are presented to markets at local and national

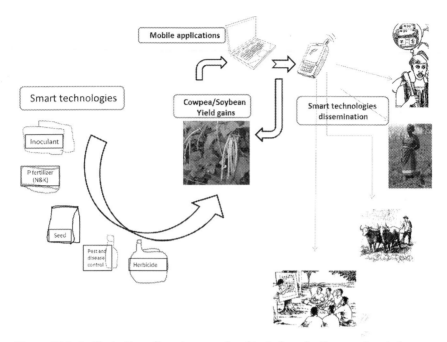

Figure 15.1 An illustration of how improved and tested production practices in legume value chains can be translated into dissemination messages, which can be scaled out using mobile phone-based information communication technology tools.

levels. The feasibility of efficient systems will be higher when the investment in rural infrastructure includes roads and rail, bulking and storage systems, and agro-input outlets.

Marketing systems also need to be more open and data based for traders to plan the flow of goods and services. A more practical example is the tobacco auctioning systems in Malawi, Zambia, and Zimbabwe. The system is based on some known numbers of contracted farmers and the opening and closing of floors are announced for all farmers with a basic floor price announced. The advantages of such a data intensive systems are clear: timeline announcements allow farmers to plan and basic floor prices are used for making tentative cash flow projections by farmers and the auctions often work for farmers who are producing more in an efficient manner. However, no system is perfect as manipulations of the highest bid prices and other malpractices have been associated with the open floor auction system. Additionally, price controls have been a setback on the part of farmers since the buyers can engage in unfair practices.

15.4 TARGETING THE MARGINAL GROUP USING FRIENDLY POLICIES

Farmers by their nature are heterogeneous; they have different levels of experience and they have variable amounts of resources available for investment into agriculture. This presents both opportunities and threats to the development and delivery of smart agricultural technologies. The major threat emanating from the cost of organizing the numerous scattered farmers throughout the whole region to make the required impact and second the majority of the small farmers use traditional and old technologies, which are not necessarily efficient and changing their mindset takes time. However, smallholder farmers are a big opportunity where change can happen and the impact of the change will be resounding. The main opportunity that national and regional agents have is that of reorganizing the farming methods and strategies that are used currently. For instance, a major policy shift can be used to influence the agricultural practices leading to a change in fortunes of many by taking a targeted approach to deliver food and business opportunities across the sector. More work is required in targeting farmers so that they can respond to climate change and variability in a more organized and effective manner.

15.5 CONCLUSIONS

Delivering smart technologies is highly feasible but it is dependent on the amount of investment governments in southern Africa will channel toward agriculture and climate change adaptation and mitigation. Climate smart agricultural systems need to be developed using the current advances in agricultural practices as the basis. Scaling out smart technologies using information and communication technologies is gaining ground and more farmers need to be engaged in these. Increasing resource-use efficiency in crop and livestock systems, reducing risk, and application of index-based insurance schemes against climate risk on agricultural value chains are major considerations in developing climate smart solutions for southern Africa.

REFERENCES

Ajanovic, A., 2011. Biofuels versus food production: does biofuels production increase food prices? Energy 36 (4), 2070–2076.
Araus, J.L., Cairns, J.E., 2014. Field high-throughput phenotyping: the new crop breeding frontier. Trends in Plant Science 19 (1), 52–61.

Batáry, P., Holzschuh, A., Orci, K.M., Samu, F., Tscharntke, T., 2012. Responses of plant, insect and spider biodiversity to local and landscape scale management intensity in cereal crops and grasslands. Agriculture, Ecosystems & Environment 146 (1), 130–136.

Chakraborty, S., Tiedemann, A.V., Teng, P.S., 2000. Climate change: potential impact on plant diseases. Environmental Pollution 108 (3), 317–326.

Fischer, G., Hizsnyik, E., Prieler, S., Shah, M., van Velthuizen, H.T., 2009. Biofuels and Food Security.

Godfray, H.C.J., Beddington, J.R., Crute, I.R., Haddad, L., Lawrence, D., Muir, J.F., Pretty, J., Robinson, S., Thomas, S.M., Toulmin, C., 2010. Food security: the challenge of feeding 9 billion people. Science 327 (5967), 812–818.

Hanjra, M.A., Qureshi, M.E., 2010. Global water crisis and future food security in an era of climate change. Food Policy 35 (5), 365–377.

Hitchcock, R.K., 2002. Coping with Uncertainty: Adaptive Responses to Drought and Livestock Disease in the Northern Kalahari. Sustainable Livelihoods in Kalahari Environments: Contributions to Global Debates, pp. 161–192.

Jarvis, A., Ramirez-Villegas, J., Campo, B.V.H., Navarro-Racines, C., 2012. Is cassava the answer to African climate change adaptation? Tropical Plant Biology 5 (1), 9–29.

Karungi, J., Adipala, E., Kyamanywa, S., Ogenga-Latigo, M.W., Oyobo, N., Jackai, L.E.N., 2000. Pest management in cowpea. Part 2. Integrating planting time, plant density and insecticide application for management of cowpea field insect pests in eastern Uganda. Crop Protection 19 (4), 237–245.

Lal, R., 2004. Soil carbon sequestration to mitigate climate change. Geoderma 123 (1), 1–22.

Marra, M., Pannell, D.J., Ghadim, A.A., 2003. The economics of risk, uncertainty and learning in the adoption of new agricultural technologies: where are we on the learning curve? Agricultural Systems 75 (2), 215–234.

Mitchell, D., 2008. A Note on Rising Food Prices. World Bank Policy Research Working Paper Series (vol.).

Nakashima, D.J., McLean, K.G., Thulstrup, H.D., Castillo, A.R., Rubis, J.T., 2012. Weathering Uncertainty: Traditional Knowledge for Climate Change Assessment and Adaptation. UNESCO and UNU, Paris and Darwin.

Naylor, R.L., Liska, A.J., Burke, M.B., Falcon, W.P., Gaskell, J.C., Rozelle, S.D., Cassman, K.G., 2007. The ripple effect: biofuels, food security, and the environment. Environment: Science and Policy for Sustainable Development 49 (9), 30–43.

Nhamo, N., Kyako, G., Dinheiro, V., 2014. Exploring options for lowland rice intensification under rainfed and irrigated ecologies in east and southern Africa: the potential application of Integrated Soil Fertility Management principles. Advances in Agronomy 128, 183–219.

Nichols, R.F.W., 1950. The brown streak disease of cassava: distribution, climatic effects and diagnostic symptoms. The East African Agricultural Journal 15 (3), 154–160.

Nkonya, E., Schroeder, T., Norman, D., 1997. Factors affecting adoption of improved maize seed and fertiliser in northern Tanzania. Journal of Agricultural Economics 48 (1–3), 1–12.

Nyong, A., Adesina, F., Elasha, B.O., 2007. The value of indigenous knowledge in climate change mitigation and adaptation strategies in the African Sahel. Mitigation and Adaptation Strategies for Global Change 12 (5), 787–797.

Paustian, K., Six, J., Elliott, E.T., Hunt, H.W., 2000. Management options for reducing CO_2 emissions from agricultural soils. Biogeochemistry 48 (1), 147–163.

Perry, B.D., Grace, D., Sones, K., 2013. Current drivers and future directions of global livestock disease dynamics. Proceedings of the National Academy of Sciences 110 (52), 20871–20877.

Rosenzweig, C., Iglesias, A., Yang, X.B., Epstein, P.R., Chivian, E., 2001. Climate change and extreme weather events; implications for food production, plant diseases, and pests. Global Change & Human Health 2 (2), 90–104.

Schmidhuber, J., Tubiello, F.N., 2007. Global food security under climate change. Proceedings of the National Academy of Sciences 104 (50), 19703–19708.

Semenov, M.A., Porter, J.R., 1995. Climatic variability and the modelling of crop yields. Agricultural and Forest Meteorology 73 (3), 265–283.

Thornton, P.K., Van de Steeg, J., Notenbaert, A., Herrero, M., 2009. The impacts of climate change on livestock and livestock systems in developing countries: a review of what we know and what we need to know. Agricultural Systems 101 (3), 113–127.

Tilman, D., Cassman, K.G., Matson, P.A., Naylor, R., Polasky, S., 2002. Agricultural sustainability and intensive production practices. Nature 418 (6898), 671–677.

Tilman, D., Socolow, R., Foley, J.A., Hill, J., Larson, E., Lynd, L., Pacala, S., Reilly, J., Searchinger, T., Somerville, C., Williams, R., 2009. Beneficial biofuels—the food, energy, and environment trilemma. Science 325 (5938), 270–271.

Tol, R.S., Van Der Grijp, N., Olsthoorn, A.A., Van Der Werff, P.E., 2003. Adapting to climate: a case study on riverine flood risks in The Netherlands. Risk Analysis 23 (3), 575–583.

Witcombe, J.R., Hollington, P.A., Howarth, C.J., Reader, S., Steele, K.A., 2008. Breeding for abiotic stresses for sustainable agriculture. Philosophical Transactions of the Royal Society of London B: Biological Sciences 363 (1492), 703–716.

FURTHER READING

Koizumi, T., 2014. Biofuels and food security. In: Biofuels and Food Security. Springer International Publishing, pp. 103–121.

INDEX

Printed in the United States
By Bookmasters